History of Mechanism and Machine Science

Volume 21

For further volumes:
http://www.springer.com/series/7481

Ruxu Du · Longhan Xie

The Mechanics of Mechanical Watches and Clocks

Springer

Ruxu Du
The Chinese University of Hong Kong
Hong Kong
People's Republic of China

Longhan Xie
South China University of Technology
Guangzhou
People's Republic of China

Additional material to this book can be downloaded from http://extras.springer.com/

ISSN 1875-3442 ISSN 1875-3426 (electronic)
ISBN 978-3-642-29307-8 ISBN 978-3-642-29308-5 (eBook)
DOI 10.1007/978-3-642-29308-5
Springer Heidelberg New York Dordrecht London

Library of Congress Control Number: 2012936760

Printed on acid-free paper

Springer is part of Springer Science+Business Media (www.springer.com)

This book is dedicated to my wife, Hui Yang,
Without her love, care and support, it would not be possible

Ruxu Du

To my wife, Wei Lin
For her understanding, support and unconditional love

Longhan Xie

Preface

Invented some 600 years ago, the mechanical watch and clock still fascinates millions of people old and young around the world today. Arguably, the mechanical watch is the most complex micro mechanical device that represents the climax of mechanical engineering and craftsmanship. When we encounter a mechanical watch, we try to look at its motion and listen to its sound. For those with an engineering background and/or a deeper interest, a question will then follow: How do they work? In the past 300 years, there have been a number of books and many hundreds of articles written about all aspects of the mechanical watch and clock. However, few examine the mechanics of the mechanical watch and clock in the context of modern mechanics.

Following the rules of economics, public interest generates business opportunities and profits. It is estimated that the mechanical watch and clock industry is a 10 billion US dollars per year industry. It is dominated by Switzerland, followed by Japan and China. In the past few years, we were fortunate to work on the design and manufacture of the mechanical watch and hence, had learnt the mechanical watch and clock in great detail. We wish to share our understanding with the readers of this monograph. It should be pointed out that the book is not for the design of the mechanical watch and clock, but for its mechanics. Also, the book is written not for the general public but for those who possess a proper background. Specifically, the required background knowledge includes calculus and mechanics at the university level.

Looking back, we remember the difficulties of working on the design and manufacture of the mechanical watch. One of the difficulties was understanding its mechanics. To overcome this difficulty, much effort has been spent. In some sense, the book is a collection of three Ph.D. theses, two M.Phil. theses and a number of B.S. Final Year Project reports and several papers from our Postdoctoral Associates. Following is a partial list of contributors:

- Dr. FU, Yu (currently with Microtechne Research and Development Center Ltd., Hong Kong SAR);

- Dr. KO, P. H. Billy (currently with Microtechne Research and Development Center Ltd., Hong Kong SAR)
- Miss TAM, L. C. Jenifer (currently with Heng Seng Bank, Hong Kong SAR)
- Miss SU, Shuang (currently Ph.D. student in the University of Pittsburg, USA)
- Prof. XU, Gu (McMaster University, Canada)
- Mr. LEI, M. C. Michael (currently M.Phil. student in the Chinese University of Hong Kong)
- Dr. XU, Wujiao (Chongqing University)
- Miss Zhiwei Li (the Chinese University of Hong Kong).

It is a joy to understand such a complex device and to make it work. We wish to share this joy with the reader.

Hong Kong, April 2012

Ruxu Du
Longhan Xie

Contents

Chapter 1
Introduction

Keeping time is an essential activity for our daily life; and it is almost as old as human civilization. Throughout history, people have invented many ways to keep time. A simple way is to examine the shadow of the sun. If you stand in the northern hemisphere on a sunny day and raise your left hand in the morning or your right hand in the afternoon while facing North, then depending on the day of the year, the shadow of your hand can be converted to the time of the day. This technique is referred to as analemmatic sundial and is illustrated in Fig. 1.1 (Wikipedia 2007).

The idea of using the shadow of the sun to read time was first used to make the obelisk (3500 B.C.E.) as shown in Fig. 1.2 (Wikipedia 2011a). The first author saw two original obelisks, one in Paris and one in Istanbul. The large Egyptian characters on these obelisks symbolized the pride of ancient Egypt.

The smaller sundial, as shown in Fig. 1.3, was found in many parts of the world. It is at least as old as the Egyptian New Kingdom period (1500 B.C.E.). There are some variations in the later days, but the principle remains the same.

Of course, the sundial does not work at night. Therefore, people invented astrolabes (Wikipedia 2011b). As shown in Fig. 1.4, it calculates the time based on stars. Such a device was considered as the climax of science in the Middle Ages and early Renaissance period.

Another simple way of keeping time is to use a water or sand funnel. However, because the force of gravity is constant, as the amount of sand decreases, the sand-dropping speed changes and will cause inaccuracy. Therefore, it only works for a fixed time period on average. Fig. 1.5 shows a Greek water clock made around 325 B.C.E. According to records, the Chinese water clock is more ancient.

The other practical problem of the water/sand clock is the water and sand itself. Water may induce the growth of bacteria and algae, whereas the viscosity of sand may change when the weather changes. These problems could easily jeopardize the accuracy of keeping time. As a result, a more reliable solution, the mechanical watch and clock, was born.

R. Du and L. Xie, *The Mechanics of Mechanical Watches and Clocks*,
History of Mechanism and Machine Science 21, DOI: 10.1007/978-3-642-29308-5_1,

Fig. 1.1 Analemmatic sundial at Saint-Etienne, France in which the user's head forms the gnomon of the dial (Wikipedia 2007)

Fig. 1.2 The Egyptian obelisk made around 3500 B.C.E. (Wikipedia 2011a)

Some 600 years ago, the mechanical watch and clock was invented. This part of the history will be briefly reviewed in the next chapter. From a mechanical point of view, the mechanical watch and clock is a device that keeps the time using only mechanical means. As a result, it is crucial to understand its mechanics and this book is written for just that purpose.

Fig. 1.3 The sundial made around 1500 B.C.E.

Fig. 1.4 An astrolabe made around 1400 A.D. (Wikipedia 2011b)

Fig. 1.5 A Greek water tower made around 325 B.C.E.

In general, the study of the mechanics includes the kinematics and the dynamics. The former concerns the motion. The mechanical watch and clock uses a unique motion mechanism called the escapement. In Chap. 2, various escapements are presented in chronological order. The latter concerns the force and vibration. This topic is discussed in Chaps. 3 and 4. The other special mechanism of the mechanical watch is the automatic winding mechanism, whose kinematics will be discussed in Chap. 5. The force of the main gear train, which is crucial as the watch and clock must run day and night without stopping, is investigated in Chap. 6. Finally, Chap. 7 gives some concluding remarks.

References

Wikipedia (2007) Analemmatic sundial. http://en.wikipedia.org/wiki/Analemmatic_sundial. Accessed 12 Dec 2011

Wikipedia (2011a) Obelisk. http://en.wikipedia.org/wiki/Obelisk. Accessed 12 Dec 2011

Wikipedia (2011b) Astrolabe. http://en.wikipedia.org/wiki/Astrolabe. Accessed 12 Dec 2011

Chapter 2
A Brief Review of the Mechanics of Watch and Clock

According to literature, the first mechanical clock appeared in the middle of the fourteenth century. For more than 600 years, it had been worked on by many people, including Galileo, Hooke and Huygens. Needless to say, there have been many ingenious inventions that transcend time. Even with the dominance of the quartz watch today, the mechanical watch and clock still fascinates millions of people around the, world and its production continues to grow. It is estimated that the world annual production of the mechanical watch and clock is at least 10 billion USD per year and growing. Therefore, studying the mechanical watch and clock is not only of scientific value but also has an economic incentive. Nevertheless, this book is not about the design and manufacturing of the mechanical watch and clock. Instead, it concerns only the mechanics of the mechanical watch and clock.

Generally speaking, a mechanical watch is made of five parts as shown in Fig. 2.1. They are the winding mechanism, the power storage, the gear train, the display and the escapement.

There are two kinds of winding mechanisms: manual winding and automatic winding. The latter is usually applied to watches and will be discussed in Chap. 5. The winding mechanism provides kinetic energy to drive the watch and clock. This energy is stored in the power storage (the mainspring). The energy from wounded mainspring drives a gear train, which usually consists of three sets of gears: the second pinion and wheel, the third pinion and wheel as well as the escape pinion and wheel. For timekeeping though, the brain is the escapement. It is the most important and most distinctive part of the mechanical watch and clock.

Why the escapement? It is well known that a watch and clock must have a precise and reliable means for timekeeping. One way is using a pendulum. It is said that pendulum theory was inspired by the swinging motion of a chandelier in the Pisa Cathedral, as shown in Fig. 2.2 (Wikipedia 2004a). Galileo Galilee (Wikipedia 2001a), Fig. 2.3, discovered the crucial property of the pendulum in 1606, which led to his decision to build a functioning pendulum clock.

R. Du and L. Xie, *The Mechanics of Mechanical Watches and Clocks*,
History of Mechanism and Machine Science 21, DOI: 10.1007/978-3-642-29308-5_2,

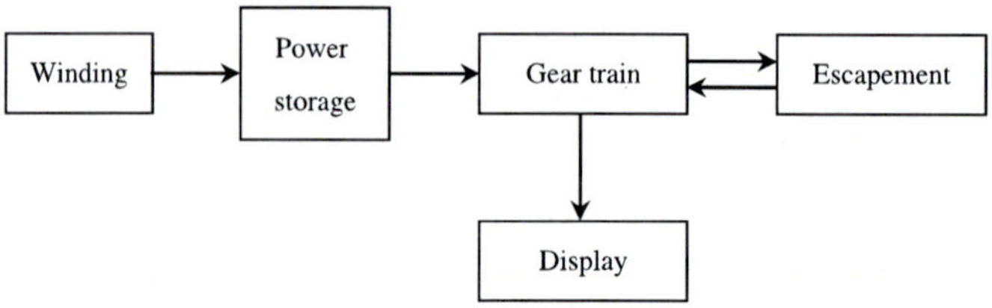

Fig. 2.1 The basic structure of mechanical watch and clock

Fig. 2.2 The chandelier in the Pisa Cathedral (Wikipedia 2004a)

Despite the fact that the swinging of the pendulum is independent of the amplitude of the swing as well as the weight of the bob, and hence, is a good means for timekeeping, it will inevitably slow down because of air resistance and mechanical friction. As a result, energy must be added, which can be done by the lift weight and/or the wound spring. However, one may imagine that as energy is gradually being used up, the driving force is gradually reducing and hence, the clock will slow down. On the other hand, when energy is being added, the clock will move faster. To solve this problem, the concept of escapement was invented. The idea is to release energy by intermittent pulses. Note that it is the impulse that drives the clock and hence, the amount of energy input does not matter. This makes the

Fig. 2.3 Portrait of Galileo Galilei (1564–1642) (Wikipedia 2001a)

clock less dependent on the stored energy and hence, much more accurate. Because such an impulsive motion is to allow the stored energy to "escape" pulse by pulse, it is, therefore, called the "escapement."

According to literature, since the time of Galileo more than one hundred different types of escapements had been created. In the subsequent sections, we discuss a number of representative escapements chronologically.

2.1 The Verge Escapement

Appearing as early as the fourteenth century, the Verge escapement is perhaps the oldest escapement (Wikipedia 2004b). It is not clear who invented this escapement, but it was certainly inspired by the alarum tower. By the sixteenth century, the working principle of the Verge escapement was well documented by the Muslim scientist Taqi al-Din Ibn Maruf (1550).

The Verge escapement is also called as the Crown-wheel-and-verge escapement. As shown in Fig. 2.4, it consists of a crown-shaped escape wheel rotating about the horizontal axis and a vertical verge. The escape wheel is driven by a lifted weight or a mainspring. Note that there are two pallets on the vertical verge shaft that are arranged at an angle. As the escape wheel rotates, one of its saw-tooth-shaped teeth turns a pallet and drives the vertical shaft in one direction (Fig. 2.4a). This also puts the other pallet in position to catch the tooth of the escape wheel on the other side. As the escape wheel continues to rotate, it drives

(a)

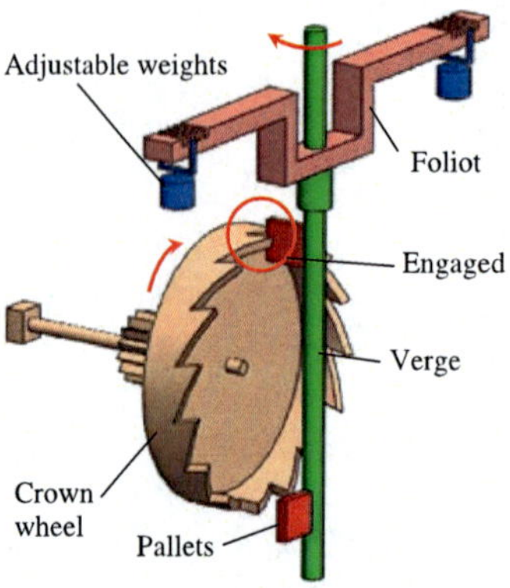

- Driven by the mainspring, the escape wheel rotates counter clockwise and catches the upper pallet in the verge
- The verge rotates clockwise

(b)

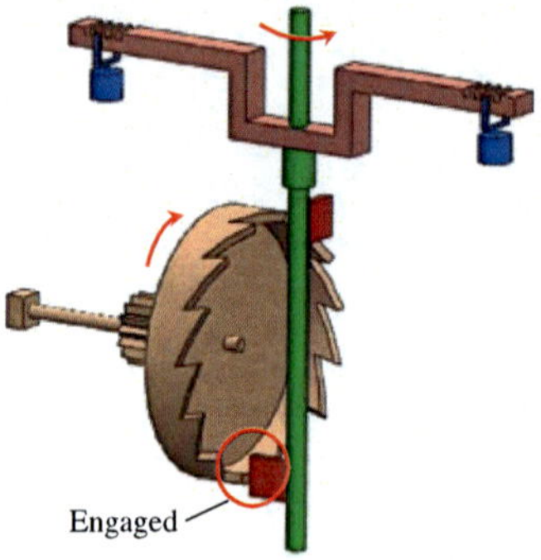

- The lower verge engages the escape wheel
- The escape wheel drives the verge rotating counter clockwise

Fig. 2.4 The operation of the Verge Escapement

the vertical shaft to rotate in the opposite direction, completing a cycle (Fig. 2.4b). The cycle then repeats converting the rotary motion of the escape wheel to the oscillating motion of the verge. Each cycle advances the wheel train of the clock moving the hands forward at a constant rate.

The Verge escapement was first used as a clock escapement and then modified into a watch escapement. Figure 2.5 shows the Verge watch escapement. From the figure, it can be seen that the crown-shaped escape wheel and the vertical shaft are the same; but the horizontal bar is replaced by a balance wheel with a hairspring. In this case, the timekeeping is regulated in part by the hairspring as it controls the engagement of the second pallet. A computer animation is shown on the Springer Website http://extra.springer.com/2012/978-3-642-29307-8.

Figure 2.6 shows a Verge escapement clock made in late 1700s (Institute of Precision Engineering 2008). The Verge escapement is usually not very accurate. This is due mainly to the fact that the driving power dominates the swinging of the verge wheel. As the driving power is consumed over time, the escapement slows down. Therefore, as new and better designs emerged, the verge escapement gradually disappeared in 1800s.

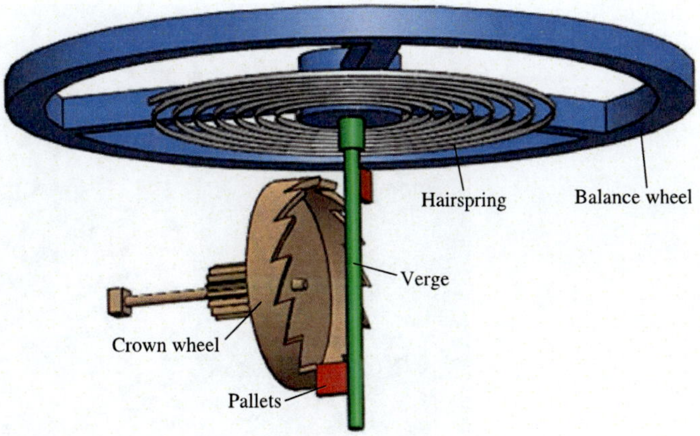

Fig. 2.5 The model of the Verge escapement watch

Fig. 2.6 A Verge escapement clock made in late 1700s (Institute of Precision Engineering 2008)

Fig. 2.7 Portrait of Robert Hooke (1635–1703) (Wikipedia 2002a)

2.2 The Anchor Escapement

The Anchor escapement is another milestone invention. It was perhaps invented by the famous British scientist Robert Hooke (1635–1703) around 1657, as shown in Fig. 2.7 (Wikipedia 2002a), and first made by the British clock master Thomas Tompion (1639–1713). However, like many of his other works, his ownership is disputed (Wikipedia 2004c). In any case, Hooke's milestone contribution to mechanical watch and clock making is indisputable.

Figure 2.8 shows the model of the Anchor escapement. It consists of an escape wheel, an anchor and a pendulum. The exact shapes of both the escape tooth and the anchor pallet are not crucial. The escape wheel is driven by a lifted weight or a wound mainspring rotating clockwise. As a tooth of the escape wheel slides on the surface of the left pallet of the anchor, the anchor moves away releasing the tooth and allowing the escape wheel to advance. Next, the pendulum reaches its highest position and swings back. It carries the right pallet towards the escape wheel, pushing the escape wheel backwards for a small distance. This locks the escape wheel until the pendulum reverses direction and the pallet begins to move away from the escape wheel, with the tooth sliding off along its surface. Then, the escape

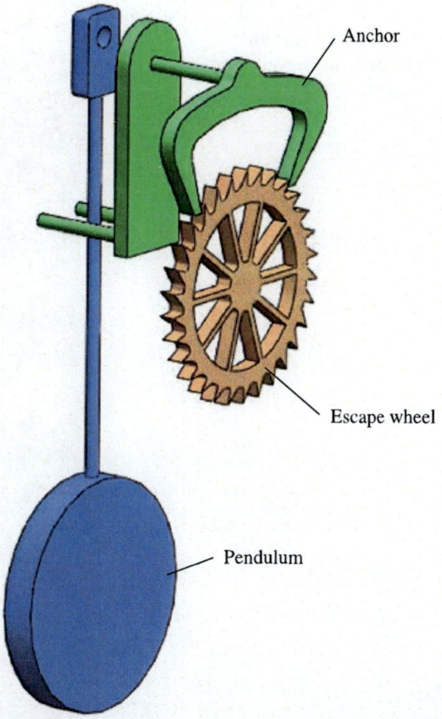

Fig. 2.8 The model of an Anchor escapement

wheel catches the left pallet again, starting a new cycle. This operation is rather similar to the Graham escapement detailed in the subsequent section.

In comparison to the Verge escapement, the Anchor escapement has two significant advantages. First, all the motions are in the same plane making the motion more stable and the manufacturing easier. Second, the pendulum needs to swing only a small angle. As discovered by Christiaan Huygens (1629–1695) (Wikipedia 2002b), this is very important because it avoids the nonlinearity present when the pendulum swings in large angles. By the early 1800s, the Anchor escapement had replaced the Verge escapement as the choice for time keeping (Fig. 2.9).

It should be mentioned that the anchor escapement had one major problem: as the escape wheel is pushed backwards, the entire gear train must move backwards and suffer from backlash. This problem is referred to as recoil and motivated many subsequent improvements, some of which will be discussed in the subsequent sections.

As for Robert Hooke, his most significant contribution to the mechanical watch and clock was not the anchor escapement but the introduction of the balance spring, also called the hairspring. Together with the clock master Thomas Tompion, who was considered the father of British clock making, the hairspring makes the mechanical watch possible. Furthermore, it gave birth to the Hooke's Law that we all learn in elementary school.

Fig. 2.9 Portrait of Christiaan Huygens (1629–1695) (Wikipedia 2002b)

2.3 Graham Escapement

In 1715, English watchmaker George Graham (1673–1751) invented the Graham escapement (Wikipedia 2004d). Born in Hethersgill, England, Graham was one of the most well-known horologists of the eighteenth century. He started his apprenticeship to an English clockmaker named Henry Aske at a young age. Later, he became the protégé of Thomas Tompion (and married his niece) and partner for life. He brought the watch and clock technology to new heights. Besides the Graham escapement, he was also the inventor of the mercury compensation pendulum, the Cylinder escapement for watches and the first chronograph. The mercury pendulum can achieve an accuracy of within a few seconds per day, a monumental achievement for the time. Graham refused to patent these inventions because he felt that they should be used by other watchmakers as well. He was a truly talented and generous inventor (Fig. 2.10).

Fig. 2.10 Portrait of George Graham (1673–1751) (Wikipedia 2004d)

The Graham escapement is also called the Deadbeat escapement. It is a modified version of the Anchor escapement and mostly eliminates the aforementioned coil problem. Figure 2.11 shows the model of the Graham escapement. Similar to the Anchor escapement, it mainly consists of an escape wheel, a pallet fork and a pendulum. During the operation, the escape wheel is driven by the power train and moves clockwise. On the other hand, the pallet fork and the pendulum are joined together and swung. Graham made a number of delicate modifications. First, the anchor pallet is concentric to its center. Second, the tip of each limb of the pallet has a specific shape designed to provide an impulse as the escape wheel tooth slides across the surface. The surface that the escape tooth strikes is called the

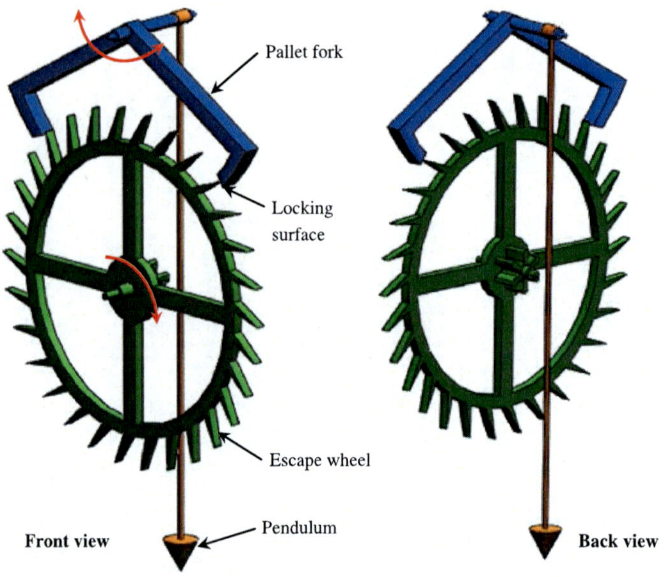

Fig. 2.11 The model of the Graham escapement

locking face, because it prevents the escape wheel from rotating farther. The design of the escape wheel, on the other hand, is relatively simple. It is made of a number of straight teeth leaning towards the direction of rotation.

The working principle of the Graham escapement is similar to that of the Anchor escapement. Figure 2.12 shows the five steps that the Graham escapement goes through in a complete cycle. Note that the circle in the figure indicates the contacting point. Figure 2.12a shows the first step, in which a tooth of the escape wheel pushes the entry pallet (the left pallet). Note that the strike point is at the locking face above the tip surface of the pallet to ensure proper contact. Both the escape wheel and the pendulum are moving forward. Figure 2.12b shows the second step, where the pendulum reaches its farthest point and begins to swing backwards. Figure 2.12c shows the third step where the exit pallet (the right pallet) of the pallet fork locks another tooth of the escape wheel and stops its motion. Its position is just right so that it does not push the escape wheel backwards, which is why it is called the Deadbeat escapement. Figure 2.12d shows the fourth step when the pendulum reaches the opposing farthest point and starts to swing forward. This unlocks the escape wheel and thus, the escape wheel can move forward again. Finally, Fig. 2.12e shows the fifth step where the pallet fork and pendulum return to their original position completing a cycle. A computer animation is shown on the Springer Website http://extra.springer.com/2012/978-3-642-29307-8.

After nearly 300 years, the Graham escapement is still used today. Figure 2.13 shows a Graham escapement clock. This kind of clock is often called the grandfather clock and has become a symbol of accuracy and reliability.

(a)

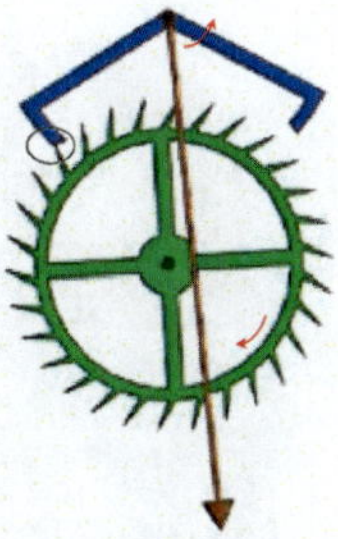

- The escape wheel moves forward. A tooth of the escape wheel impacts the entry pallet, giving the first shock;
- The pallet fork and the pendulum swing counter clockwise;
- The escape wheel continues to move forward.

(b)

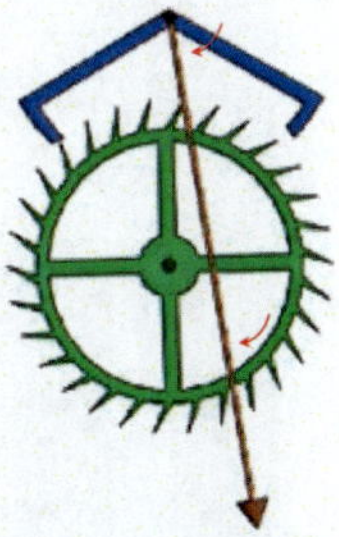

- The pendulum reaches its highest point and starts to swing backwards;
- Following the pendulum, the pallet fork swings clockwise;
- The escape wheel continues to move forward;
- No shock actually occurs.

(c)

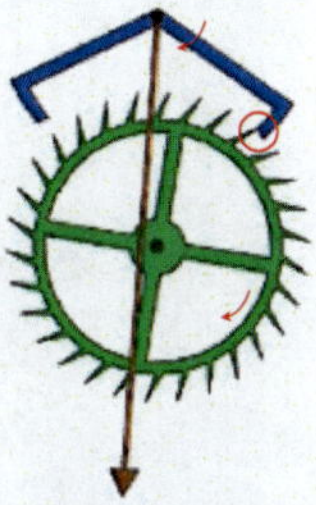

- The pendulum and the pallet fork swing clockwise;
- The exit pallet locks another tooth of the escape wheel, causing another shock and stops the motion of the escape wheel;
- The pendulum continues to swing clockwise.

(d)

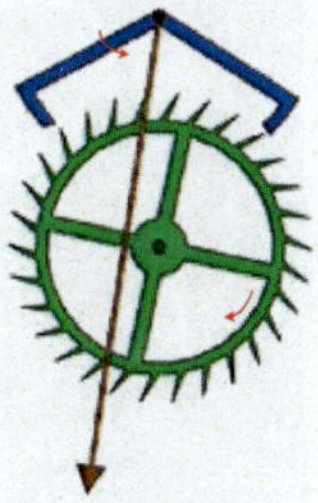

- The pendulum reaches the opposing highest point and starts to swing backwards;
- Following the pendulum, the pallet fork swings counter clockwise;
- The escape wheel continues to move forward;
- No actual shock occurs.

(e)

- The pendulum and the pallet fork swing counter clockwise;
- The escape wheel continues to advance, reaching the position in the first step and finishing the cycle.

Fig. 2.12 The operation of the Graham escapementGraham escapement

Fig. 2.13 A modern clock based on Graham escapement

Fig. 2.14 Portrait of John Harrison (1693–1776) holding his trophy H4 watch (Wikipedia 2001b)

2.4 The Grasshopper Escapement

The Grasshopper escapement was invented by another famous English clockmaker, John Harrison (1693–1776) around the year 1722 (Wikipedia 2001b, 2004e). During this time, England was gaining power at sea. Evidently, estimating the position on the sea was extremely important. It was known that latitudinal position could be estimated based on astronomical charts; but longitudinal position was a challenge for maritime navigation. Isaac Newton (1643–1727) argued that astronomical positioning could also be used, but an easier method for accurate timekeeping was based on reference to the home base. The solution for establishing the longitude of a ship at sea was considered so intractable that the British Parliament offered a prize of £20,000 (about 4.6 million USD today), called the Longitude Prize, for an accurate clock. John Harrison devoted his life's work to building such a clock.

As mentioned in the previous sections, by then a number of different types of escapements had been invented. Modified from the Anchor escapement, the Grasshopper escapement was used in his first three maritime time keepers: Harrison Number One (H1) through Harrison Number Three (H3). He then spent another 29 years on the project and finally won the Longitude Prize. Although the grasshopper escapement was not used in his final Harrison Number 4 (H4), which was a watch, it left a mark in history (Fig. 2.14).

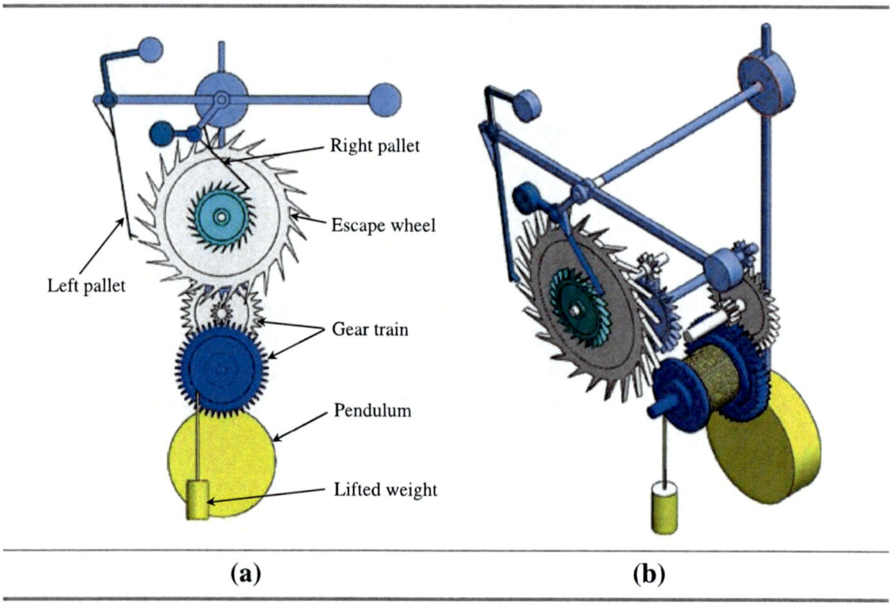

Fig. 2.15 Illustration of the Grasshopper escapement. **a** Front view **b** Back view

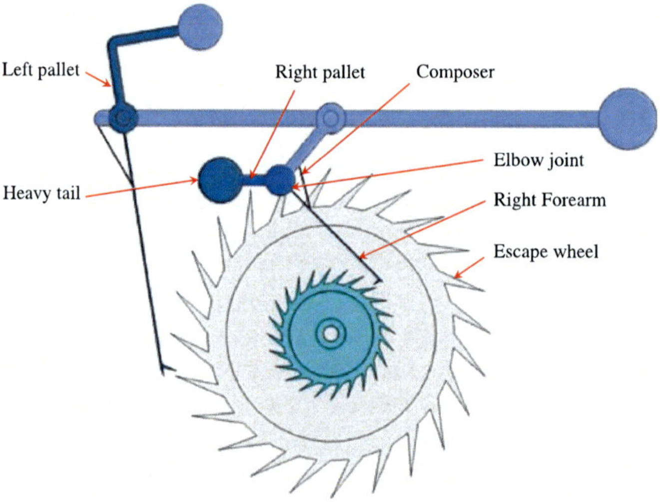

Fig. 2.16 The two pallets and the escape wheel in the Grasshopper escapement

The Grasshopper escapement was also evolved from the anchor escapement. It has been suggested that the name of this escapement comes from the resemblance of the pallet arms to the legs of a grasshopper. As shown in Fig. 2.15, the escapement consists of an escape wheel, a pendulum, a driving mechanism (the lifted weight) and two pallets shaped like a grasshopper.

Figure 2.16 gives the details of the escapement: The right pallet has an elbow joint connected to a heavy tail and a forearm, as well as a composer. The tail is slightly heavier so that the forearm tends to move away from the escape wheel. The composer prevents the forearm from rising further. As the upper arm rotates clockwise, the tip of the pallet at the forearm is pushed downwards. When lifted by the escape wheel, the pallet will take the composer with it, and when released, it will return to the resting position. The left pallet has a similar structure. The two pallets are joined with the pendulum on the upper arm.

The operation of the Grasshopper escapement is shown in Fig. 2.17. In Fig. 2.17a, the gear train is turning the escape wheel clockwise and the pendulum is swinging to the left. This will lift the left pallet. The impulse comes to an end when the right pallet, which is moving down towards the escape wheel, catches a tooth of the escape wheel by its notch. In Fig. 2.17b, the right pallet tries to reverse the clockwise motion of the escape wheel. However, the pendulum is still swinging to the left and the upper arms must always move with it. Thus, the right elbow recoils, crumpling a little in the process. As soon as this happens, the right pallet becomes separated from its tooth and hops away, finding itself free to revert to its natural angle. It might have also been this "hop" that gave the escapement its nickname. In Fig. 2.17c, as the pendulum returns from its left extreme, the escape wheel continues its clockwise motion and its impulse through the right pallet. In Fig. 2.17d, the escape wheel is recoiled by the arrival of the left pallet, causing the right pallet to hop away. A computer animation is shown on the Springer Website http://extra.springer.com/2012/978-3-642-29307-8.

The Grasshopper escapement has two advantages: its regularity of operation and its absence of sliding friction. It does not recoil like the Anchor escapement. Rather, one pallet is released only by the engagement of the other, as shown in Fig. 2.17b and d. The impulse given to the pendulum is thus uniform in both its amount and timing. In addition, it does not have much friction. This is because the forearm helps the pallets to jump in and out of the escape wheel teeth. Also, there is little sliding friction. Although there remains friction on the pivots, it is minor in comparison to that of the sliding in the anchor escapement.

There are, however, several limitations that made the Grasshopper escapement uncompetitive. First, when the pallets are in contact with the escape wheel, the drive to the escape wheel is interrupted, and when the drive is restored, the escape wheel may accelerate rapidly and uncontrollably. Second, when the power runs down, the pallets have a tendency to be unable to return to their proper stop positions. Third, and most importantly, even more than other clock escapements, the Grasshopper escapement pushes the pendulum back and forth throughout its cycle. The pendulum is never allowed to swing freely. This disturbs the pendulum's natural motion as a harmonic oscillator and causes a lack of isochronism.

Although the Grasshopper escapement was never popular, it was used by several famous clocks. The most famous one is John Harrison's H1 (Wikipedia 2001c) built between 1730 and 1735, as shown in Fig. 2.18. Two hundred and seventy years later, in 2008, as a tribute to the Grasshopper escapement, the Corpus Clock was built at Corpus Christi College, Cambridge University, in Cambridge, England. It is shown in Fig. 2.19 (Wikipedia 2008). The ever moving grasshopper exemplifies that the time is another dimension of the universe.

(a)

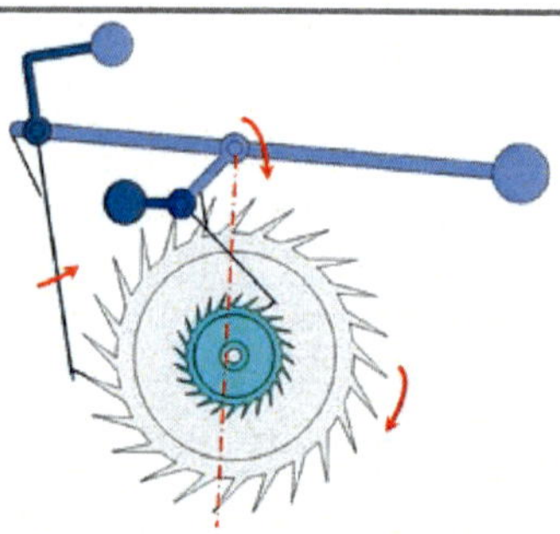

- The escape wheel advances;
- The pendulum swings clockwise;
- The left pallet is lifted;
- The right pallet catches a tooth of the inner escape wheel, causing a shock.

(b)

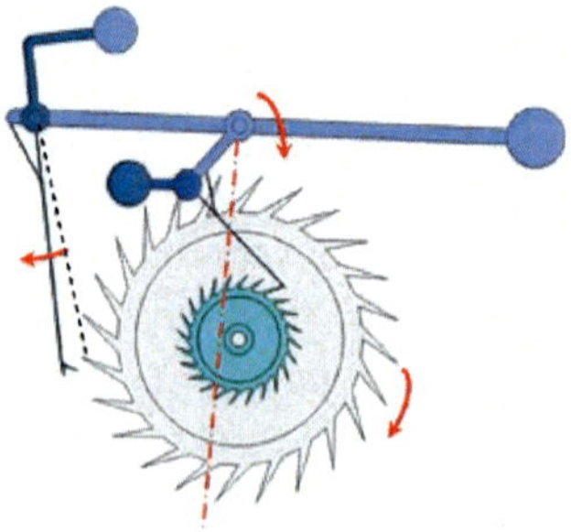

- The right pallet tries to push the escape wheel backwards;
- The pendulum continues to swing clockwise, causing the right elbow to recoil;
- The right pallet 'hops', releasing the escape wheel;
- The escape wheel advances.

(c)

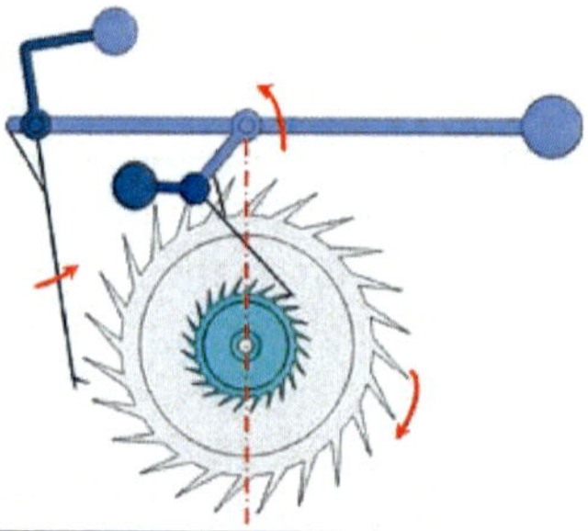

- The pendulum reaches its right extreme and starts to swing backwards;
- The left pallet moves counter clockwise;
- The escape wheel advances and one of its external teeth catches the left pallet, causing another shock.

(d)

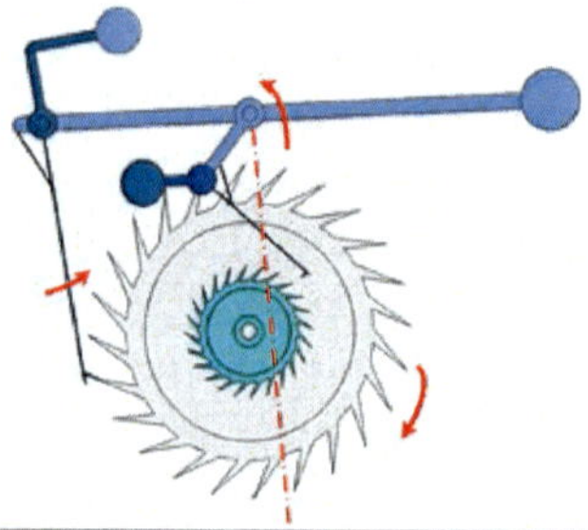

- Similar to the second step, the left pallet ;hops', releasing the escape wheel;
- The escape wheel continues to advance;
- The pendulum swings counterclockwise until reaching its left extreme and then starts to swing clockwise.

Fig. 2.17 The operation of the Grasshopper escapement

Fig. 2.18 John Harrison's H1 built between 1730 and 1735 (Wikipedia 2001c)

2.5 The Spring Detent Escapement

The spring detent escapement, most commonly used on some nineteenth century's precision watches, is a type of detached escapement. The early form was invented by the French watchmaker Pierre Le Roy (1717–1785) in 1748 (Fig. 2.20), who created a pivoted detent type of escapement (Wikipedia 2004f). It was then generalized in 1783 by the English watchmaker Thomas Earnshaw (1749–1829) (Wikipedia 2005) with his standard spring detent escapement and used until mechanical chronometers became antediluvian (Fig. 2.21). Although John Arnold (1736–1799) and Swiss watchmaker Ferdinand Berthoud (1727–1807) both had their own design in 1779, neither of their designs could match Earnshaw's design in popularity. Due to the virtual absence of sliding friction between the escape tooth and the pallet during impulse, the spring detent escapement could be made more accurate than lever escapements. Unfortunately, the spring detent escape was rather fragile, not self-starting and harder to manufacture in volume. In 1805, Earnshaw and Arnold's son (by then John Arnold was deceased) were awarded by the Board of Longitude for their contributions to chronometers. Earnshaw was also known for his bimetallic temperature compensator, and Arnold simplified the complicated structure of the chronometer by applying a helical balance spring.

Fig. 2.19 The Corpus Clock at Corpus Christi College, Cambridge. (Wikipedia 2008)

The spring detent escapement is shown in Fig. 2.22. It consists of an escape wheel, a roller with an impulse pallet and a locking pallet, as well as a detent made of a blade, a horn and a spring.

The operation of the spring detent escapement is shown in Fig. 2.23. In Fig. 2.23a, before the first shock, the impulse roller rotates counter clockwise and the horn of the detent locks the escape wheel. Figure 2.23b shows the first shock. The impulse roller contacts the detent causing the first shock and pressing it down allowing the escape wheel to move forward. Figure 2.23c shows the second shock; it occurs when the escape wheel catches the impulse pallet. After the second shock in Fig. 2.23d, the impulse roller swings backwards locking the escape wheel. In Fig. 2.23e, the impulse roller reaches the spring and has the momentum to lift it up, causing the third shock, while the escape wheel continues its clockwise swinging. After the third shock in Fig. 2.23f, lifted by the discharging roller, the spring reaches its highest position. A computer animation is shown on the Springer Website http://extra.springer.com/2012/978-3-642-29307-8.

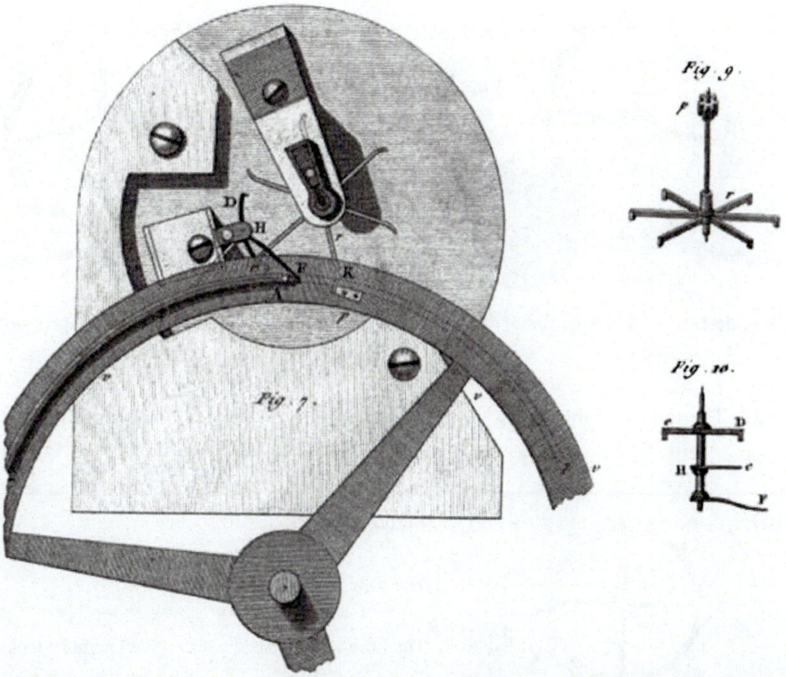

Fig. 2.20 Pierre Le Roy's detent escapement (Wikipedia 2004f)

Fig. 2.21 Portrait of Thomas Earnshaw with his chronometer (Wikipedia 2005)

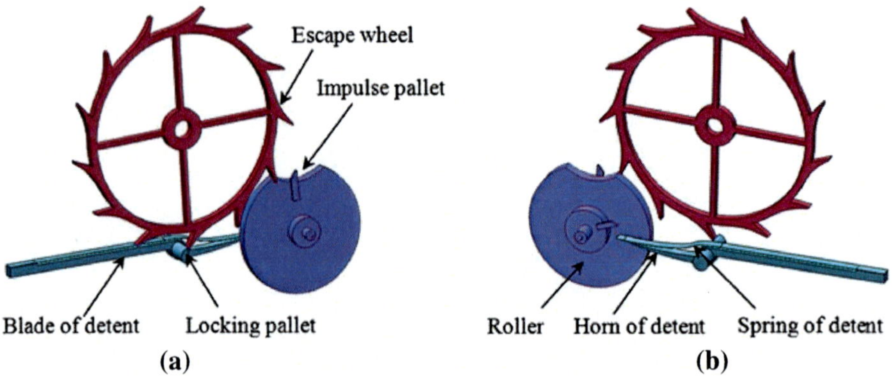

Fig. 2.22 The model of the spring detent escapement. **a** Front view **b** Back view

(a) 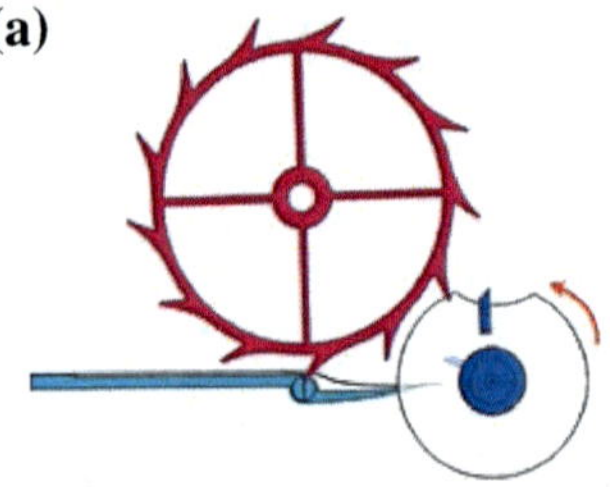	• Before 1st Shock • The roller swings counter clockwise; • The horn and the spring stay still. The horn locks the escape wheel.
(b) 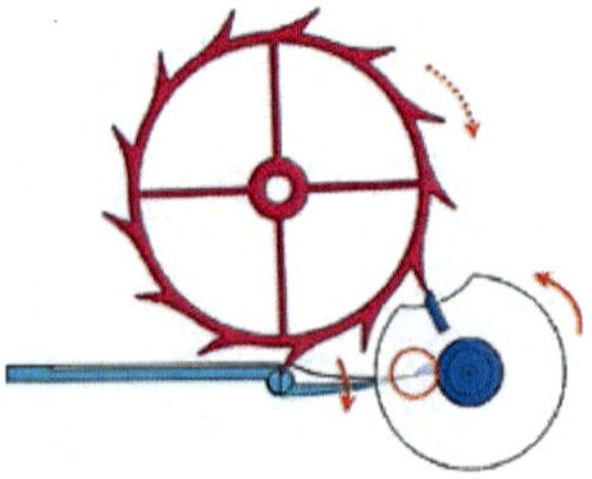	• At the 1st Shock • The impulse roller swings counter clockwise and the roller contacts the spring and presses the horn of detent causing the first half of 1st shock; • The horn of the detent swings clockwise and the locking pallet releases the escape wheel; • Driven by the gear train, the escape wheel moves forward in the clockwise direction (as the dotted line shows).
(c) 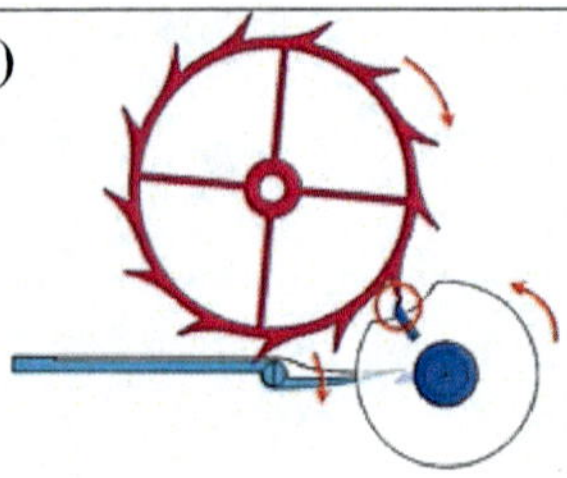	• At the 2nd Shock • The impulse roller continues to swing counter clockwise but is slowing down; • The escape wheel catches the impulse pallet, causing the 2nd Shock.

Fig. 2.23 The operation of the spring detent escapement

(d)

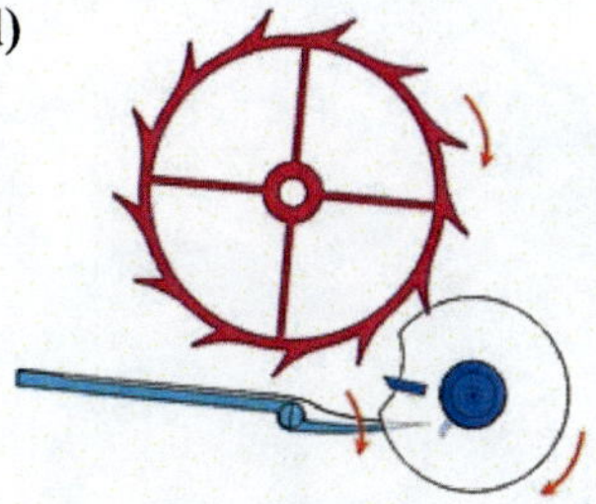

- After the 2nd Shock
- The impulse roller reverses to swing clockwise;
- The escape wheel is unlocked, swings clockwise.

(e)

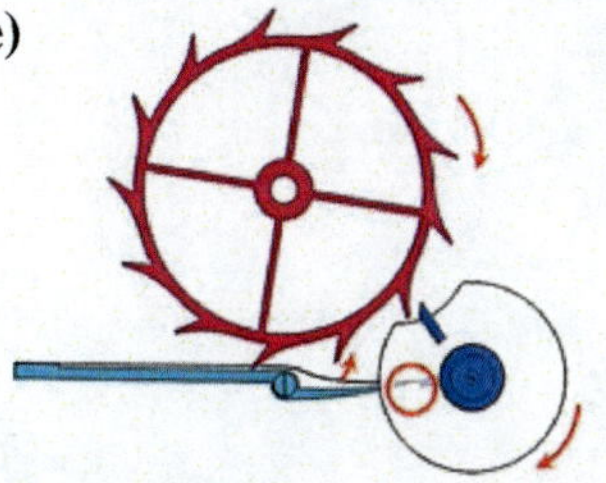

- At the 3rd Shock
- The roller reaches the spring and has the momentum to lift it up, causing the 3rd Shock.
- The impulse roller and the escape wheel continue swinging clockwise.

(f)

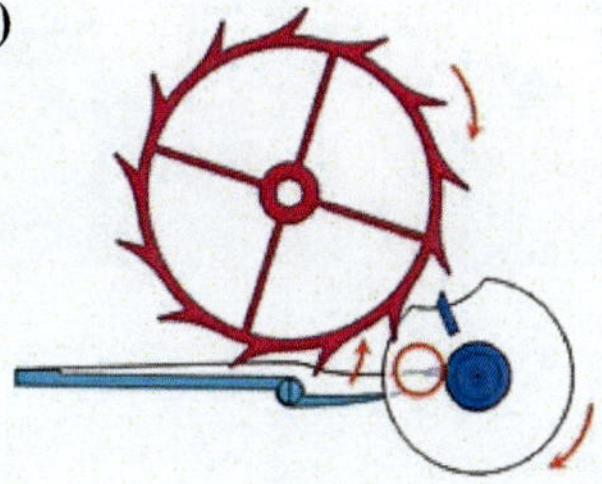

- After the 3rd Shock
- The discharging roller lifts the spring to its highest position and continues swinging clockwise until the discharging roller reaches its limit position.
- The spring will go back to its balance position after being released by the discharging roller.
- The next tooth of the escape wheel enters the lock of the locking pallet, beginning the next cycle.

Fig. 2.23 continued

The spring detent escapement uses a spring strip to regulate the timekeeping. While it is simple, its accuracy and reliability are limited. Thus, the spring detent escapement was gradually being phased out in late 1800s. Today, one may still be able to find the spring detent escapement watches in antique stores. Figure 2.24 shows a chronometer with spring detent escapement made by J. Calame Robert, in which a spiral hairspring had been added.

Fig. 2.24 A chronometer with spring detent escapement made by J. Calame Robert

2.6 The Cylinder Escapement

The cylinder escapement got its name from its cylinder-shaped balance wheel. Different from other escapements, the cylinder escapement does not have a pallet fork. It only consists of an escapement wheel and a balance wheel. From a historical point of view, the cylinder escapement was an improvement over the verge escapement. It was first used for clocks, as shown in Fig. 2.25.

Figure 2.26 shows the model of the cylinder escapement in watch. Note that the escape wheel teeth of this escapement lie in a horizontal plane, so this escapement is also known as the "horizontal escapement" (introduced by George Graham in 1726). The escape wheel usually has thirteen to fifteen wedge-shaped teeth, standing above the rim of the wheel with the pointed end of the "wedge" leading. Mounted on the balance staff is a polished steel tube or hollow cylinder. Nearly one half of which is cut away, allowing the escape wheel teeth to enter as the balance wheel swings back and forth. As each tooth enters the cylinder, it impulses the balance wheel on the entry lip of the cylinder wall. The tooth rests within the cylinder while the balance wheel completes its oscillation and begins its return journey. In due course, the tooth escapes from within the cylinder, again giving impulse as it leaves. The succeeding tooth, which has been held against the outside wall of the cylinder while the first tooth is within, then enters the cylinder and the process is repeated. Later, the cylinder shell was made out of ruby, which was

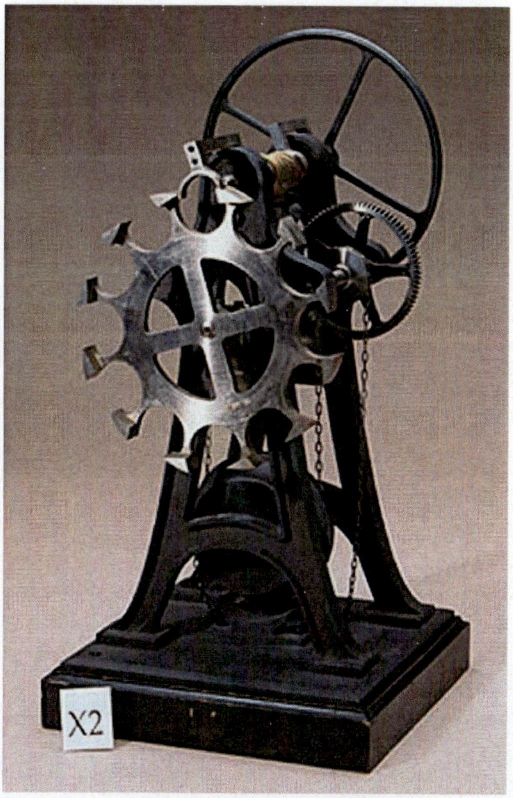

Fig. 2.25 A cylinder escapement clock

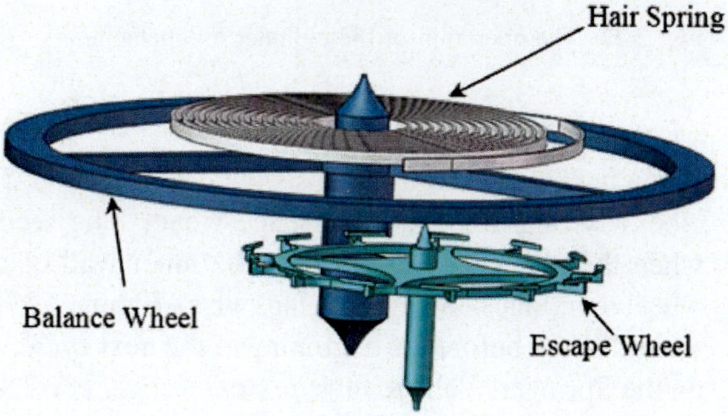

Fig. 2.26 The CAD model of the cylinder escapement

probably introduced by John Arnold in 1764, though not used to large extent. John Ellicott also used it in his later watches, and it became popular with Breguet for a period of time.

The operation of the cylinder escapement is shown in Fig. 2.27. Figure 2.27a shows that the escape wheel is about to be released. Figure 2.27b shows the first shock. It happens between the escape wheel and the outer wall of the cylinder, and the escape wheel is locked after then. After the first shock, as shown in Fig. 2.27c, the escape wheel is locked and the balance wheel swings to its limit position and about to

(a)

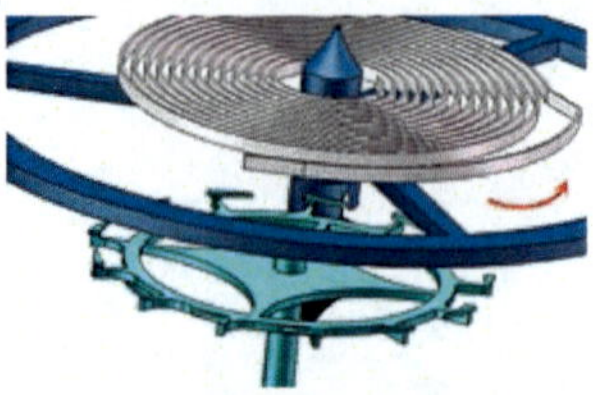

- Before the 1st Shock
- The balance wheel swings counter clockwise;
- The escape wheel is locked by the inner wall of the cylinder and is about to unlock and swing clockwise;

(b)

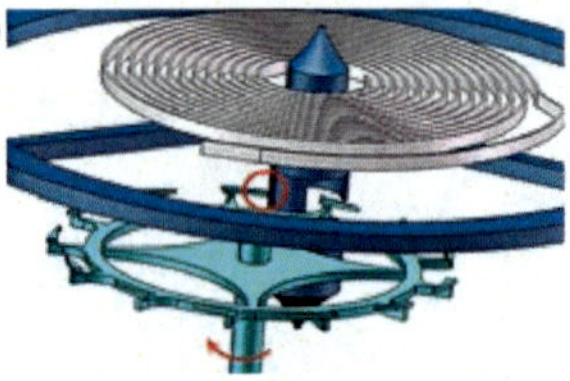

- The 1st Shock
- The balance wheel continues to swing counter clockwise and its hollow tube unlocks the escape wheel;
- The escape wheel swings clockwise until locked by the outer wall of the cylinder, causing the 1st shock;

(c)

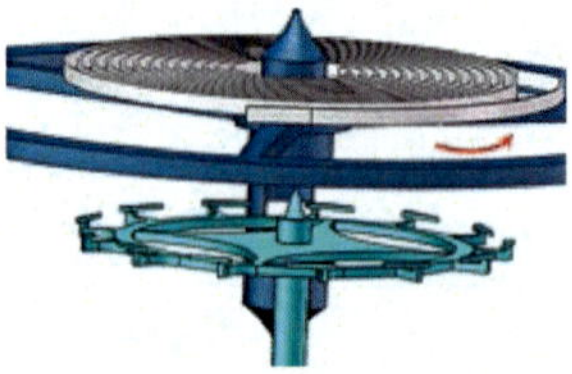

- After the 1st Shock
- The escape wheel is locked by the cylinder and remain stopped;
- The balance wheel reaches its limit position and then is about to swing back in clockwise direction;

Fig. 2.27 The operation of the cylinder escapement

swing backwards. Before the second shock in Fig. 2.27d, the balance wheel swings clockwise and unlocks the escape wheel. The second shock occurs in Fig. 2.27e when the escape wheel contacts the inner wall of the cylinder, while the balance wheel continues swinging clockwise. Figure 2.27f shows movements after the second shock before the beginning of the next cycle. A computer animation is shown on the Springer Website http://extra.springer.com/2012/978-3-642-29307-8.

The constant contact between the balance wheel and the escape wheel makes the cylinder escapement prone to wear and sensitive to dirt and hence, needs regular cleaning. Besides, it is hard to manufacture and maintain. Therefore, it was gradually being replaced in late 1800s. Today, one may still be able to find watches made of cylinder escapement in antique stores, like the one shown in Fig. 2.28.

(d) 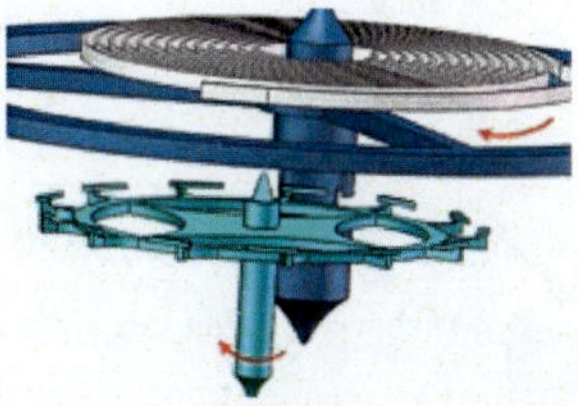

- Before the 2nd Shock
- The balance wheel swings clockwise and unlocks the escape wheel;
- The escape wheel swings clockwise ;

(e) 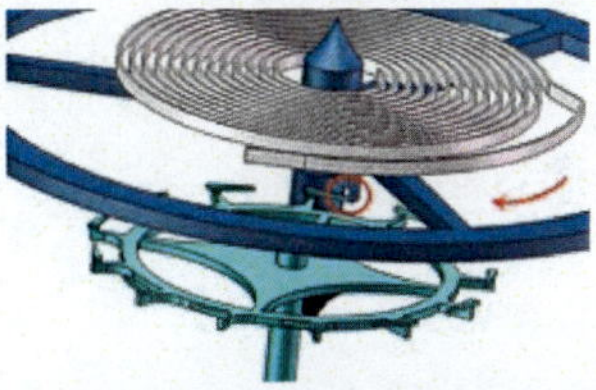

- The 2nd Shock
- The escape wheel contacts the inner wall of the cylinder causing the 2nd shock;
- The balance wheel continues to swing clockwise

(f) 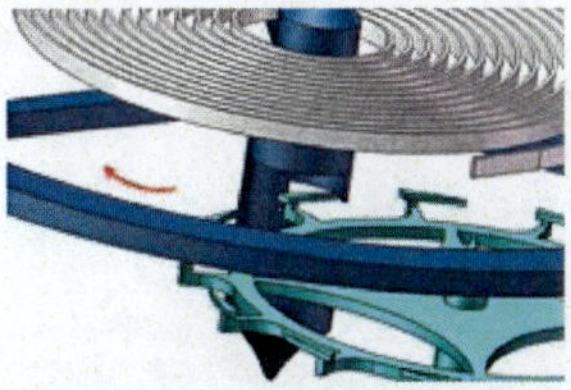

- After the 2nd Shock
- The balance wheel swings to its limit position and is about to swing backward in counter clockwise direction;
- The escape wheel is locked waiting for the next cycle.

Fig. 2.27 continued

Fig. 2.28 Zylinderwerk 2 using cylinder escapement

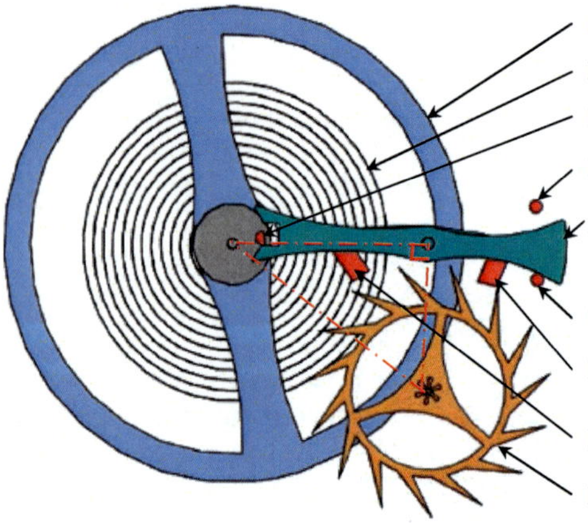

Fig. 2.29 The model of the English lever escapement

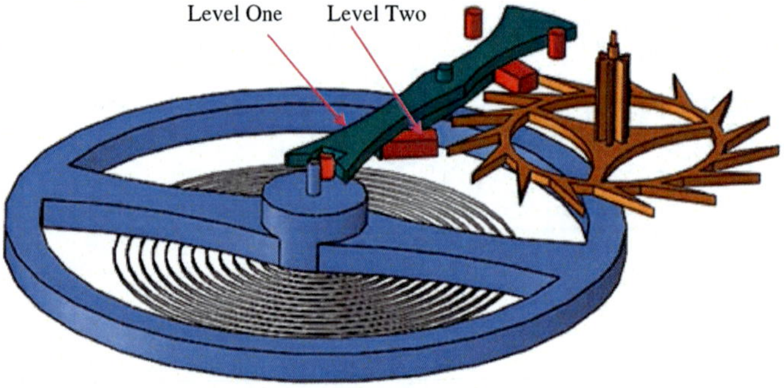

Fig. 2.30 The two levels of pallet fork in the English lever escapement

2.7 The English Lever Escapement

This escapement was invented by English clockmaker Thomas Mudge (1715–1794) in 1754 and hence, was referred to as the English escapement. Mudge was apprenticed to George Graham. It is one of the earliest escapements that does not require a pendulum and hence, is regarded as a milestone in the history of watch and clock.

As shown in Fig. 2.29, the English lever escapement is composed of four parts: the escape wheel, the anchor-like pallet fork, the balance wheel and the hairspring. The axes of the balance wheel, the pallet fork and the escape wheel form a right-angled triangle (as shown by the dot-dash line).

Note that the pallet fork has two levels as shown in Fig. 2.30: Level 1 is the balance wheel level, on which the balance wheel has a half-cycle shaped ruby, called the impulse pin, to turn the pallet fork. Level 2 is the escape wheel level, on which the escape wheel turns the two rubies on the pallet fork, called the entry pallet jewel and the exit pallet jewel. There are also two position pins that limit the

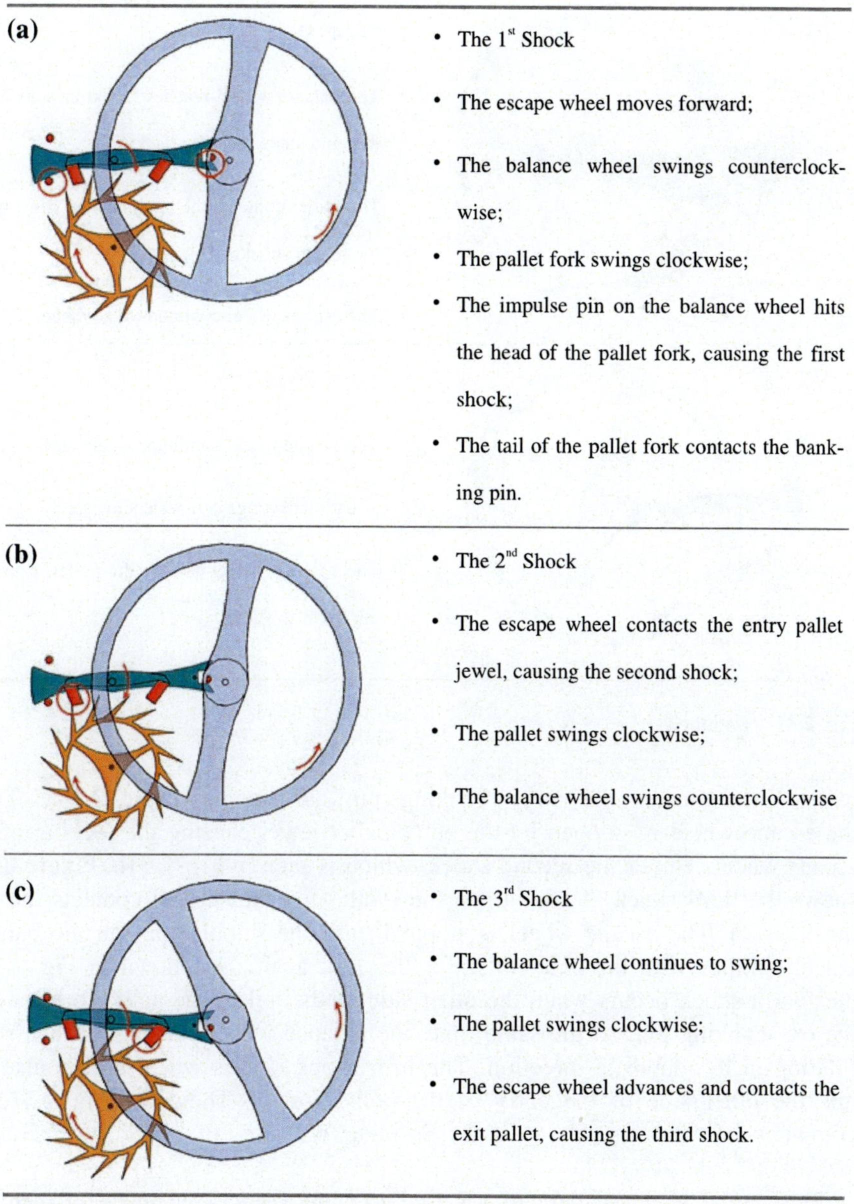

Fig. 2.31 The operation of the English lever escapement

swing of the pallet fork. The big innovation of the English lever escapement is the use of the hairspring. It allows the balance wheel swinging in a large angle and hence, is much more reliable.

Figure 2.31 shows the operation of the English lever escapement in a cycle. It is made of five shocks. The first shock is the contact of the impulse pin on the balance wheel with the head of the pallet fork. At the meantime, the tail of the

(d)

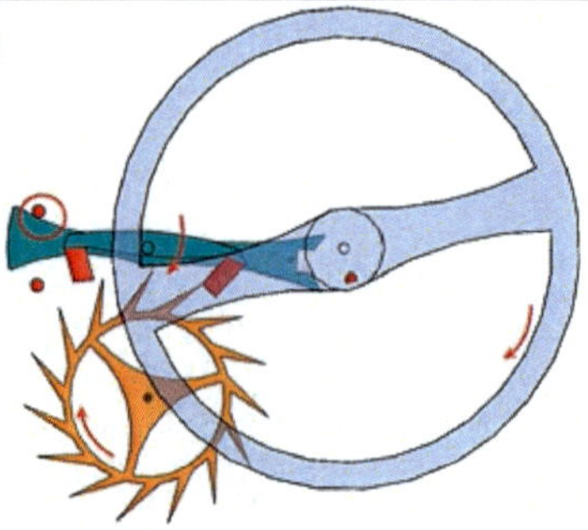

- The 4th Shock
- The balance wheel reaches its extreme location and starts to swing back;
- The pallet contacts the banking pin, causing the fourth shock;
- The escape wheel continues to advance.

(e)

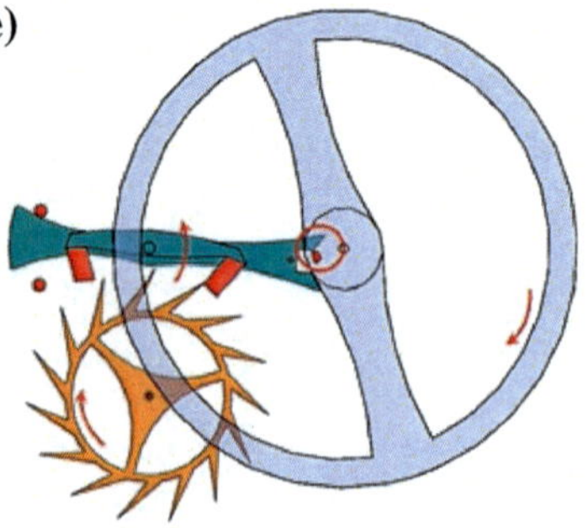

- The 5th Shock
- The escape wheel continues to advance;
- The pallet swings counterclockwise;
- The balance wheel catches the pallet, causing the fifth shock.

Fig. 2.31 continued

pallet fork touches the entry banking pin as shown in Fig. 2.31a. Then, the escape wheel is stopped by the entry pallet jewel causing the locking of the escape wheel. This is the second shock, which is seen in Fig. 2.31b. Figure 2.31c shows the third shock. At this time, the tooth of escape wheel contacts the exit pallet jewel. The escape wheel is stopped until the impulse pin on the balance wheel collides with the entry of the pallet fork again. As shown in Fig. 2.31d, the fourth shock occurs when the other side of the tail of the pallet fork touches the exit banking pin. At the same time, the balance wheel pauses and then starts rotating in the opposite direction. The fifth shock occurs when the impulse pin hits the other side of the entry of the pallet fork as shown in Fig. 2.31e. A computer animation is shown on the Springer Website http://extra.springer.com/2012/978-3-642-29307-8.

The English lever escapement was used for many years in the nineteenth century. One may still find them in antiques today, such as the one shown in Fig. 2.32.

2.8 The Swiss Lever Escapement

Appearing in the middle of the nineteenth century in Switzerland, the Swiss lever escapement is a modification of the English lever escapement. It is not clear who invented the Swiss lever escapement, though it was probably a team effort. It has

Fig. 2.32 An antique pocket watch with the English lever escapement

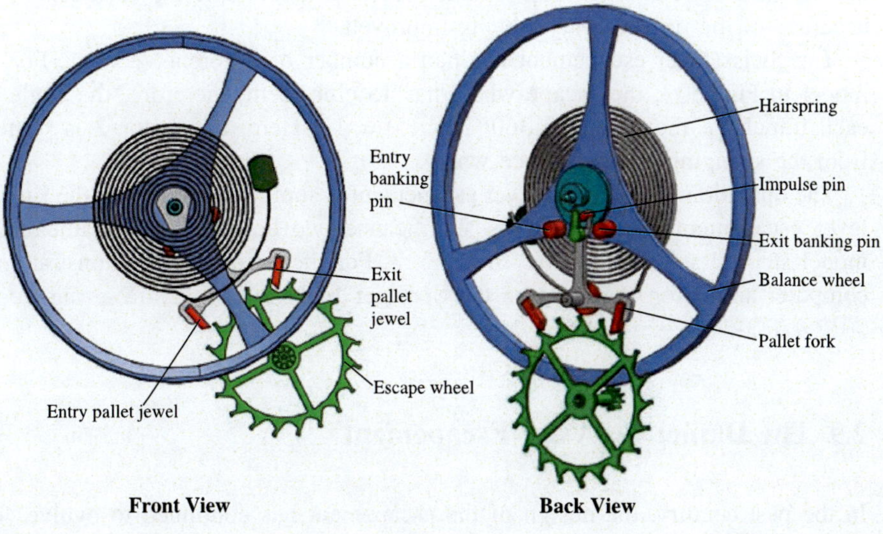

Fig. 2.33 The CAD model of the Swiss lever escapement

been the most commonly used escapement in the world ever since. In fact, at least 98% of the existing mechanical movements use this escapement because of its high degree of accuracy and reliability.

A model of the Swiss lever escapement is shown in Fig. 2.33. Similar to the English escapement, it consists of a balance wheel, hairspring, pallet fork and escape wheel. The pallet is shaped like a fork, giving it the name "pallet fork." The pallet fork results in two significant improvements over the English lever escapement. First, the centers of the balance wheel, the pallet fork and the escape wheel are aligned in one line, making the power transmission more efficient and

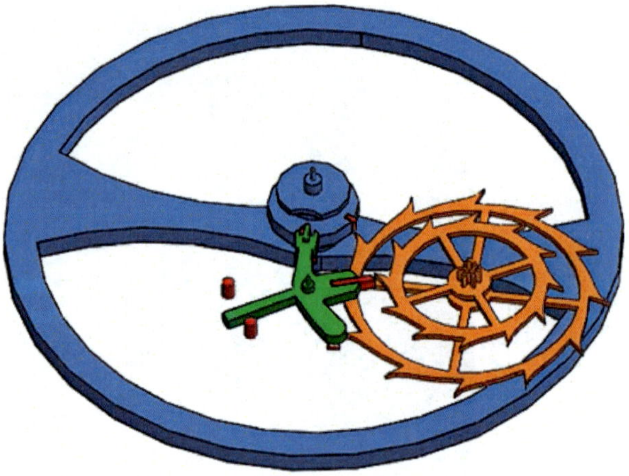

Fig. 2.34 The model of the Daniels Co-Axial Escapement

stable. Second, the pallet fork needs only to swing a small degree (around 50° and 25° on each side) and hence, does not get much nonlinearity. Consequently, the accuracy of the timekeeping is being improved.

The Swiss lever escapement has had a number of different versions. For the model in Fig. 2.33, the escape wheel has 15 club teeth; therefore, the angle for each impulsive movement is $360/(2 \times 15) = 12$. Here, the factor 2 is resulted from the swinging of the balance wheel.

The operation of the Swiss lever escapement is somewhat similar to the English lever escapement. Because of its significance, we developed its mathematical model step by step, as detailed in Chap. 3. For the purpose of demonstration, a computer animation is shown on the Springer Website http://extra.springer.com/2012/978-3-642-29307-8.

2.9 The Daniel Co-Axial Escapement

In the past century, the design of the escapement has continued to evolve. The most significant invention is perhaps the Daniels co-axial double-wheel escapement. It is the masterpiece of Dr. George Daniels (1926–) (Wikipedia 2009). Dr. Daniels is a professional horologist with many achievements. Besides inventing the co-axial escapement, he is also the author of several books on mechanical watch movement (Cecil Cluttoh and George Daniels 1979; George Daniels 1981; George Daniels 2011) and the former president of the Horological Institute.

Figure 2.34 shows the model of the Daniels co-axial escapement. It is more complicated than the Swiss lever escapement and has three levels. Figure 2.35 shows the three levels of the escapement. On Level 1, the balance wheel contacts the pallet fork. The guard pins are also on this level. The escape wheel has two levels, one for the inner escape wheel and the other for the outer escape wheel,

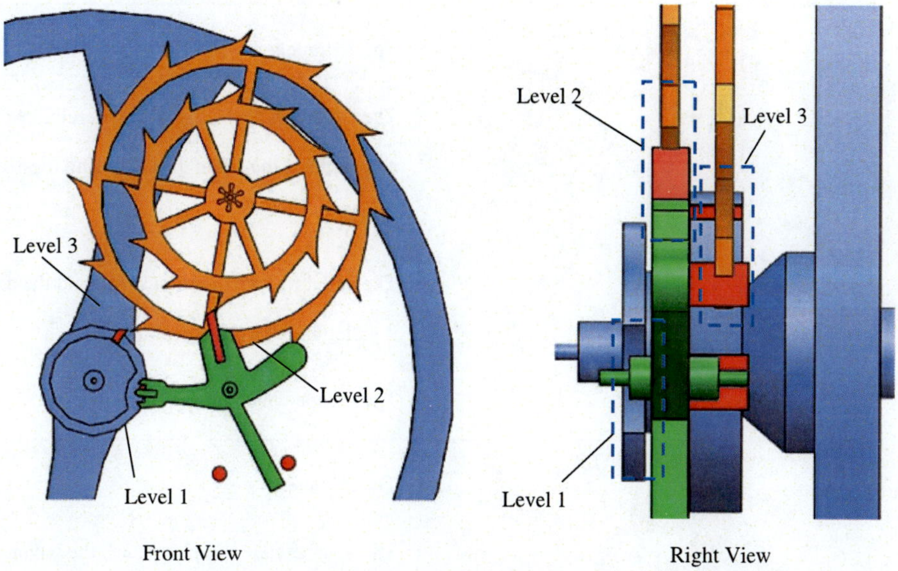

Fig. 2.35 The three levels of the Daniels co-axial escapement

with 12 teeth on each level. On Levels 2 and 3, the two levels of the pallet fork contact the two levels of the escape wheel.

As shown in Fig. 2.36, the Daniels co-axial escapement has six shocks in a cycle. Figure 2.36a shows the first shock. It occurs when the semi-circular impulse-pin on the balance contacts the entry pallet of the pallet fork to unlock the escape wheel. The escape wheel advances and the pallet fork swings clockwise. Figure 2.36b shows the second shock. It happens when the outer escape wheel shocks the impulse stone at the balance wheel. Then, the balance wheel continues to swing counter-clockwise until the escape wheel contacts the pallet fork to cause the third shocks, as shown in Fig. 2.36c. Next, the inner escape wheel touches the trapezium-shaped impulse stone of fourth shock and is shown in Fig. 2.36d. Here, the other side of the entry pallet of the pallet fork contacts the semi-circle shaped impulse-pin on the balance wheel. This impulse unlocks the pallet and escape wheel, causing them to rotate clockwise. Figure 2.36e shows the fifth shock when the inner escape wheel is stopped by the locking-stone on the pallet fork. The sixth shock, as shown in Fig. 2.36f, occurs when the pallet fork contacts the other guard pin. The cycle of the escapement is then completed. A computer animation is shown on the Springer Website http://extra.springer.com/2012/978-3-642-29307-8.

In comparison to the Swiss level escapement, the Daniels co-axial escapement has the same balance wheel and hairspring, but different escape wheel and pallet fork. These changes result in several advantages. First, within a cycle, the pallet fork only contacts the escape wheel once and hence, is more efficient. Second, the pallet fork swings only a small angle (30° and 15° on each side), thereby reducing the effect of nonlinearity. Finally, it minimizes the sliding between the escape

(a)

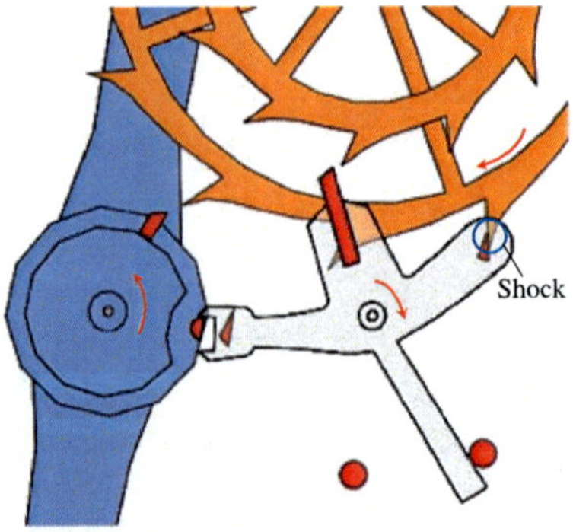

- The 1st Shock
- The balance wheel swings counterclockwise and contacts the pallet fork, causing the first shock;
- The pallet fork swings clockwise and the exit pallet unlocks the escape wheel;
- The escape wheel moves forward.

(b)

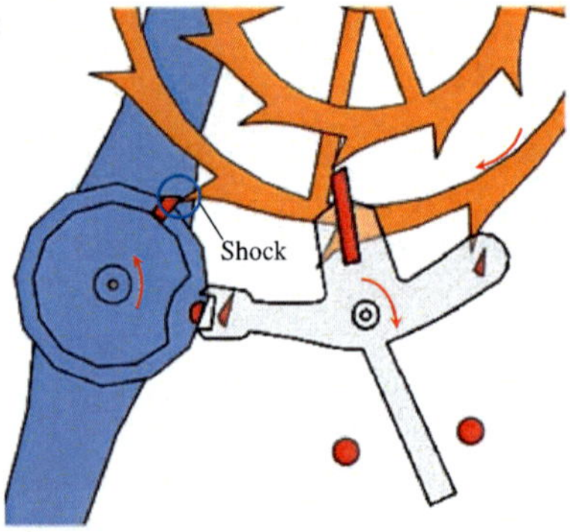

- The 2nd Shock
- The outer escape wheel contacts the balance wheel, causing the second shock;
- The balance wheel continues to swing counterclockwise;
- The pallet fork continues to swing clockwise.

(c)

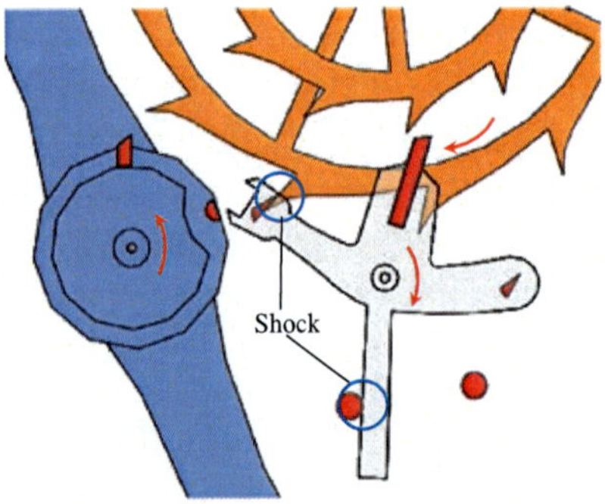

- The 3rd Shock
- The balance wheel continues to swing counterclockwise;
- The pallet fork contacts the banking pin;
- The escape wheel contacts the pallet fork, causing the third shock.

Fig. 2.36 The operation of the Daniels' co-axial escapement

(d)

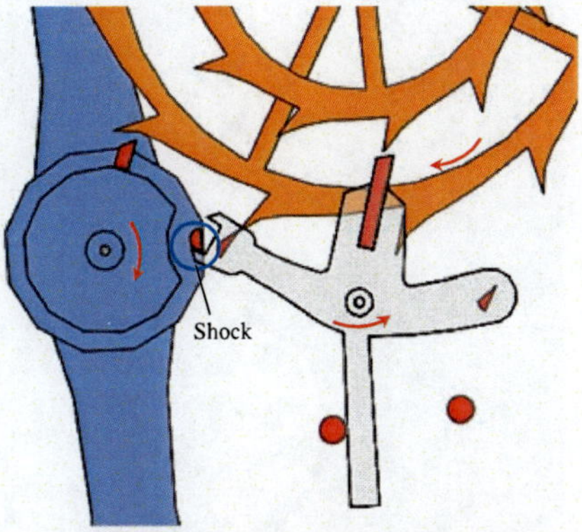

- The 4th Shock
- The balance wheel reverses to swing clockwise;
- The pallet fork contacts the balance wheel, causing the fourth shock;
- The escape wheel is unlocked.

(e)

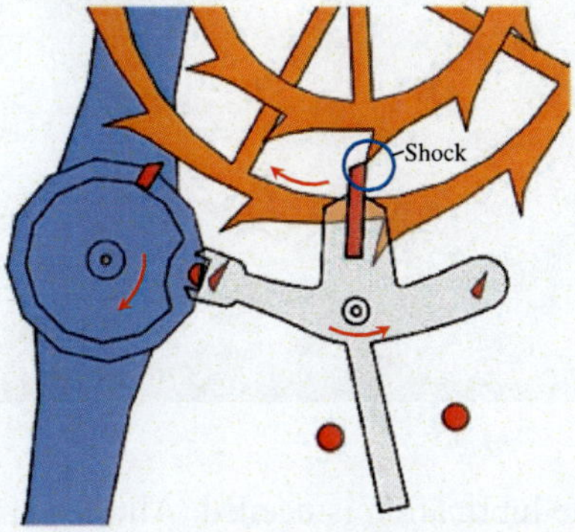

- The 5th Shock
- The pallet fork swings counterclockwise;
- The escape wheel contacts the pallet fork, causing the fifth shock;
- The balance wheel continues to swing clockwise and contacts the pallet fork, unlocking the escape wheel

(f)

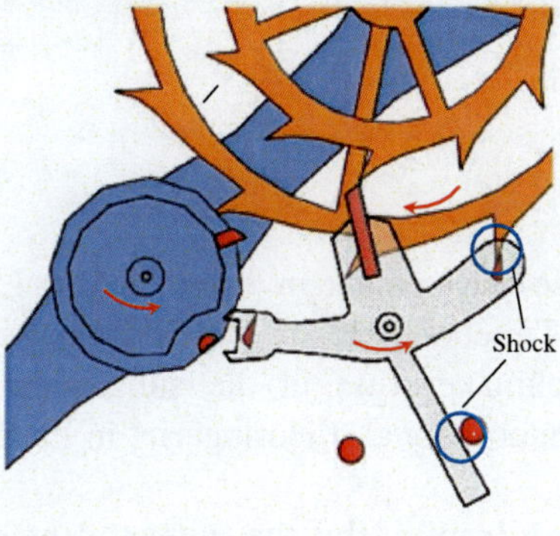

- The 6th Shock
- The pallet fork contacts the banking pin, causing the sixth shock;
- The escape wheel continues to move forward and catches the pallet fork;
- The balance wheel reaches its extreme and starts to swing counterclockwise.

Fig. 2.36 continued

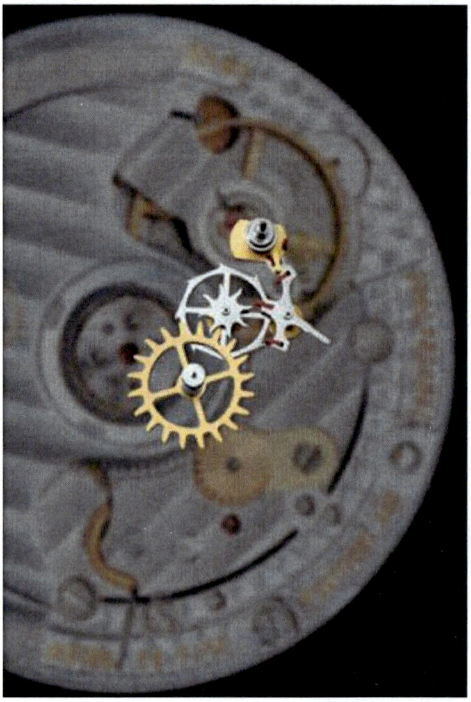

Fig. 2.37 Omega movement based on Daniels's Co-Axial Escapement

wheel and the pallet fork and hence, little lubrication is needed. All these improve the isochronisms. However, the trade-off is a more difficult manufacturing process. This escapement was adopted by Omega Co. for production since 1980s till the present day. A sample is shown in Fig. 2.37.

2.10 The Dual Ulysse Escapement

The newest escapement is the dual Ulysse escapement invented by Dr. Ludwig Oechslin in 2004 (Ludwig Oechslin 2004; Timebooth 2011). Dr. Oechslin received his Ph.D. in 1983 and his master watchmaker title in the subsequent year. Presently, he is the curator of the Musée International d'Horlogerie, in La Chaux-de-Fonds, Switzerland.

The dual Ulysse escapement is perhaps inspired by the independent double wheel escapement invented in 1800s. Figure 2.38 shows the model of the double wheel escapement. Like many old designs, the independent double wheel escapement was abandoned because of its complexity and lack of reliability.

As shown in Fig. 2.39, the dual Ulysse escapement consists of a balance wheel with a plate and a hairspring, a triangle-shape lever with two horns and two recesses and two escape wheels. There are also two pins used to limit the swing of the lever. Its most notable feature is the two escape wheels with specially designed tooth profile. Escape wheel 1 is driven by the gear train and meshes with Escape

Fig. 2.38 The independent double wheel escapement

wheel 2. The two escape wheels also interlock with each other under the control of the lever. The lever receives pulses generated alternately by the first and the second escape wheels and transmits these pulses to the plate on the balance wheel, driving the balance wheel to swing. It also locks the first and the second escape wheels alternately. Thus, the lever fulfills a dual function: transmits the force to the balance wheel and locks the escape wheels alternately.

The operation of the dual Ulysse escapement is illustrated in Fig. 2.40. Figure 2.40a shows the safety phase that occurs before the first shock. Note that the lever is in contact to both escape wheels creating the lock. At the same time, the balance wheel is reaching its limit position and is about to swing clockwise. Figure 2.40b shows the unlocking phase, at which the first shock occurs when the plate on the balance wheel contacts the upper left horn of the lever causing it pivoting counter clockwise. This releases the two escape wheels. The locking phase is shown in Fig. 2.40c, in which the second and the third shocks occur almost at the same time. First, Escape wheel 1 contacts the first recess of the lever causing the second shock. Then, the upper right horn of the lever catches the plate of the balance wheel causing the third shock. At this time, the lever is in contact to both escape wheels creating the lock. Figure 2.40d is another safety phase, in which the balance wheel continues to swing, while the lever and the two escape wheels remain locked. Figure 2.40e shows the other safety phase. It is symmetric to Fig. 2.40b. Figure 2.40f shows the other unlocking phase, which is symmetric to Fig. 2.40c. Finally, Fig. 2.40g shows the other locking phase, which is symmetric to Fig. 2.40d. After that a new cycle will start.

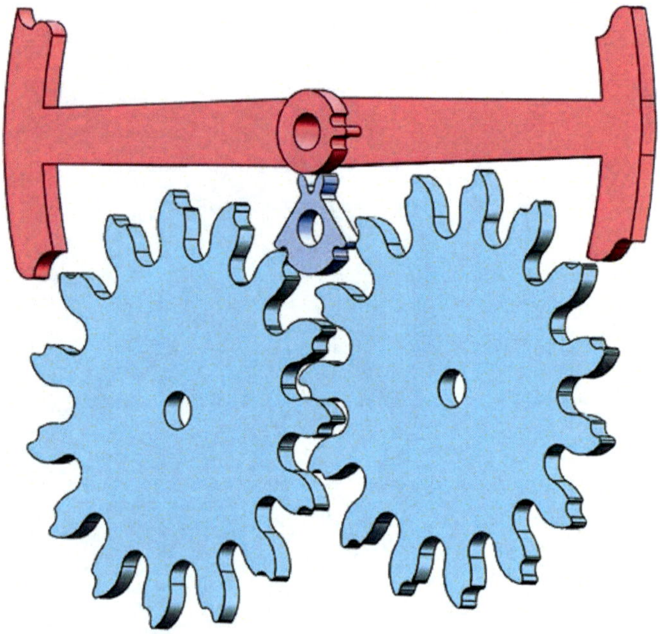

Fig. 2.39 The model of the Dual Ulysse escapement

The dual Ulysse escapement has several advantages; the symmetric escape wheel design makes the power transportation more effective. However, it also demands for higher manufacturing and assembly accuracy, as small errors may cause problems such as recoil. At present, Ulysse Nardin uses this escapement in its product line as shown in Fig. 2.41.

2.11 Concluding Remarks

According to the literature survey, there have been a number of monographs on the mechanical watch and clock. For example, George Daniels wrote two books: *Watchmaking* (George Daniels 1981) and *The Practical Watch Escapement* (George Daniels 2011). In particularly, the first one covers all the fundamentals of watch making. He also co-authored a book on the history of the watch and clock (Cecil Cluttoh and George Daniels 1979). Donald de Carle wrote a handbook, but is relatively brief (Donald de Carle 1984). The recent book by Eric Bruton is also interesting (Eric Bruton 2004).

In recent years, with the development of the Internet, many Websites were developed to demonstrate the mechanical watch and clock online. Although these Websites are usually less organized, frequently changed and often inaccurate, they outperform books in terms of accessibility, readability (videos and animations make the contents easier to understand) and popularity. Some useful Websites include

(a)

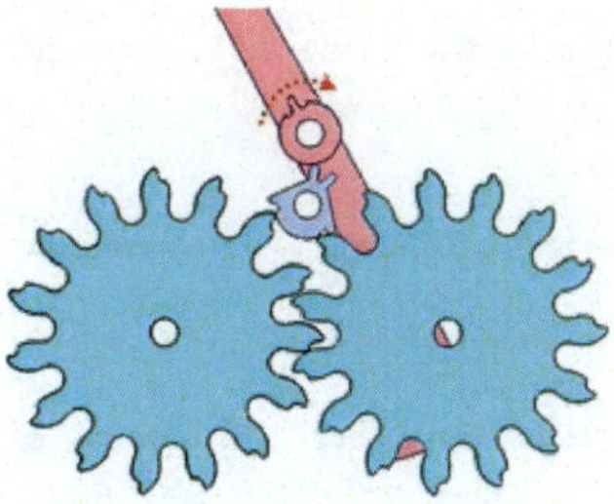

- Before the 1st Shock
- Escape wheel 1 is locked by the lower left corner of the lever;
- Escape wheel 2 is locked by escape wheel 1;
- The lever is in contact with Escape wheel 2;
- The balance wheel is now at its limit position and is about to swing clockwise;

(b)

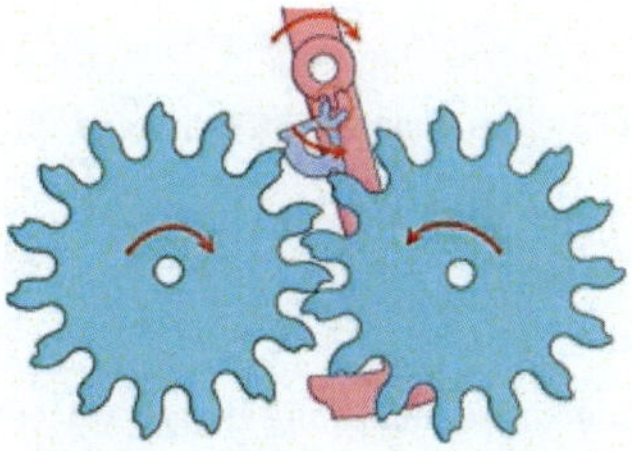

- The 1st Shock
- The plate on the balance wheel contacts the upper left horn of the lever creating the 1st Shock and causes lever pivoting counter clockwise;
- The balance wheel continues swinging clockwise;
- Escape wheel 1 is unlocked and swings clockwise;
- Escape wheel 2 meshes with escape wheel 1 and swings counter clockwise;

(c)

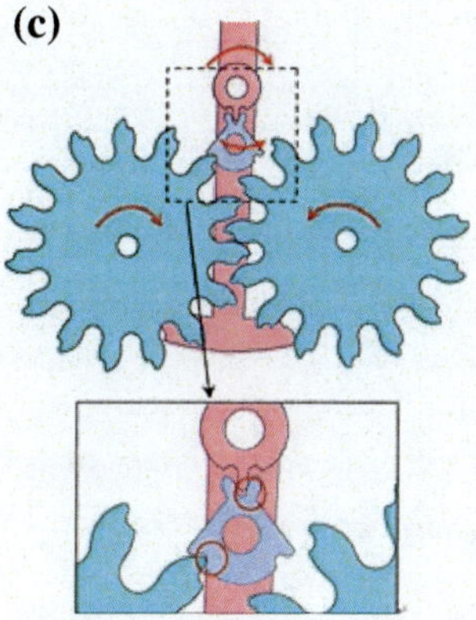

- The 2nd and 3rd Shock
- One tooth of the released Escape wheel 1 strikes the first recess causing the 2nd Shock and accelerates the lever to swing counter clockwise.
- The upper right horn of the lever catches up the plate of the balance wheel causing the 3rd Shock. This also recharges the sprung-balance system;
- The balance wheel continues to swing clockwise.

Fig. 2.40 The operation of the Dual Ulysse escapement

Clock Watch (Volker Vyskocil 2012), *Clock and Watch Escapement Mechanics* (Mark Headrick) and A *Brief History of Precision Timekeeping* (Ozdoba). In particularly, we recommend our website *Virtual Library of Mechanical Watch and Clock* (Institute of Precision Engineering, the Chinese University of Hong Kong 2011).

However, few studies have been carried out on the mechanics of the mechanical watch and clock, especially in the context of modern physics and mathematics.

(d)

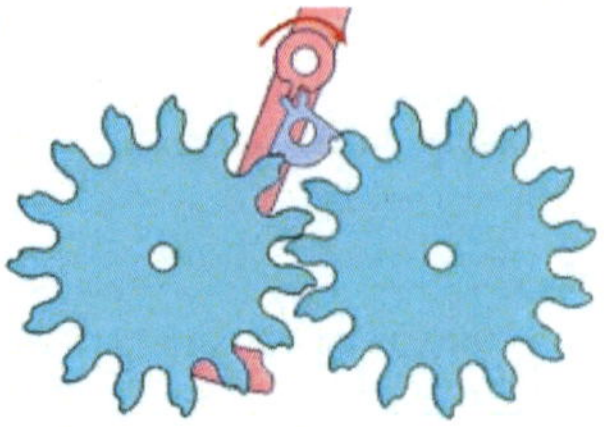

- After the 3rd Shock
- The plate on the balance wheel exits the contact with the lever;
- The lever reaches locks Escape wheel 2 by its second recess;
- Escape wheel 2 locks Escape wheel 1; Escape wheel 1 locks the lever.
- The balance wheel continues to swing clockwise

(e)

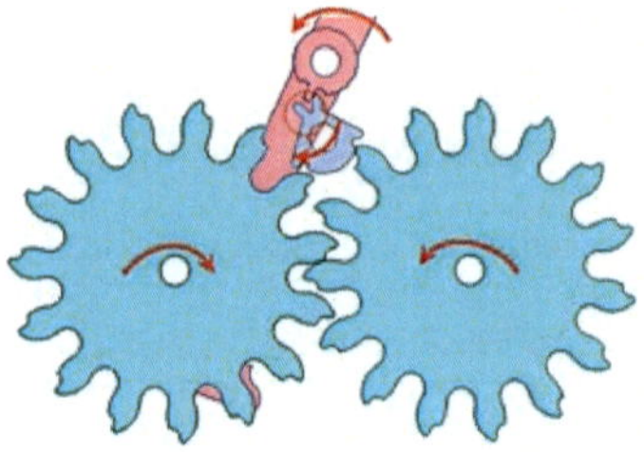

- The 4th Shock
- The balance wheel swings counter clockwise and its plate contacts the upper left horn of the lever causing the 4th Shock;
- The lever pivots clockwise, unlocking Escape wheel 2; Escape wheel 2 swings counter clockwise unlocking Escape wheel 1;
- Escape wheel 1 meshes with Escape wheel 2 and swings clockwise;

(f)

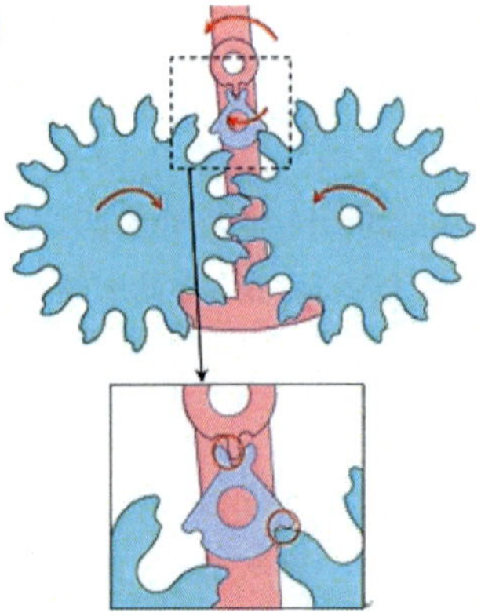

- The 5th and 6th Shock
- One tooth of the released escape wheel 2 strikes the second recess causing the 5th shock, and accelerates the lever;
- Like it is in the 3rd shock, the upper left horn of the lever catches up the plate of the balance wheel causing the 6rd shock. This also recharges the sprung-balance system.

Fig. 2.40 continued

(g)

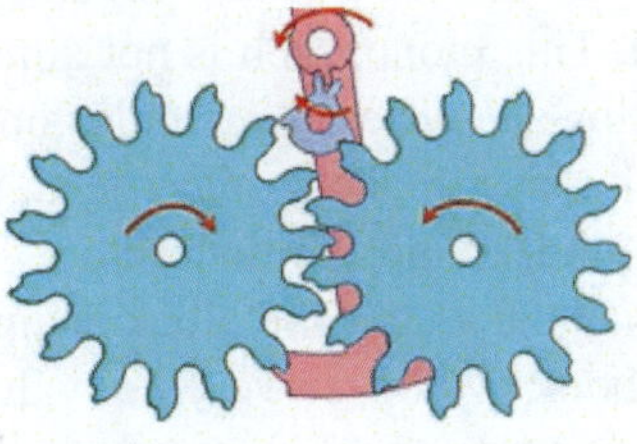

- After the 6th Shock
- The plate is about to quit the contact to the lever and continues swinging counter clockwise;
- Escape wheel 1 swings clockwise and then is locked by the lever; Escape wheel 2 swings counter clockwise and then is locked by Escape wheel 1;
- The lever is about to reach its limit position, after which the next cycle begins.

Fig. 2.40 continued

Fig. 2.41 Ulysse Nardin's "Freak Blue Phantom" with dual Ulysse Escapement

This is due in part to the fact that after the 1900s, watch making was considered a solved problem, to which scientists and engineers are less devoted. After the introduction of the quartz watch in the 1970s, modern electronics have taken the center stage. Scientists and engineers consider the mechanical watch and clock interesting, but not important. Furthermore, the mechanics of the mechanical watch and clock is rather complex. It takes a large amount of effort to simply understand its operation, not to mention its mechanics.

We had the opportunity to work on a project aimed at designing and manufacturing mechanical watch movements in 2005–2009. As a result, much information was collected and knowledge was gained. This monograph is not aimed at design and manufacture, but towards presenting a systematic study on the mechanics of the mechanical watch and clock. The readers may need solid background in engineering mathematics to fully understand the material.

It is crucial to note that the mechanics of mechanical watch and clock is unique, as is this book. Therefore, we hope that this book would serve as a valuable document for those both already in the business and interested in it. After all, the technology that has existed for more than six centuries is worth to be commended. As Elbert Einstein said, "the only reason for time is so that everything does not happen at once." To note when events occur, we need to keep time. To keep time using a machine, we need to know its mechanics.

References

Cecil Cluttoh and George Daniels (1979) Watches: a complete history of the technology and development of the watch, 3rd edition, Sotheby Parke Bernet Publishers Limited, London

Donald de Carle (1984) Watch & Clock Encyclopedia. Random House Value Publishing, New York

Eric Bruton (2004) The History of Clocks & Watches. Chartwell Books

George Daniels (1981) Watchmaking, Philip Wilson Publishers Limited, New York

George Daniels (2011) The Practical Watch Escapement, Philip Wilson Publishers Limited, New York

Institute of Precision Engineering, the Chinese University of Hong Kong (2008) Verge Escapement. http://www.ipe.cuhk.edu.hk/projects10_vl=7English.html. Accessed 12 Dec 2011

Institute of Precision Engineering, the Chinese University of Hong Kong (2011) Virtual library of mechanical watch and clock. Available at http://www.ipe.cuhk.edu.hk/projects10_library.html. Accessed 12 Dec 2011

Ludwig Oechslin (2004) Escapement for a Timekeeper. US Patent 6,708,576 B2, March 2004

Mark V. Headrick. Clock and watch escapement mechanics, Abbey Clock Clinic, Available at http://www.abbeyclock.com/tframed.html.

Ozdoba C (2011) A brief history of precision timekeeping—Part2: the 18th century. Available at: http://www.ozdoba.net/swisswatch/history_part2.html. Accessed 12 Dec 2011

Tam LC, Fu Y, Du R (2007) Virtual library of mechanical watch movements. CAD Solutions, Honolulu, 25–29

Taqi al-Din Ibn Maruf (1550) The Brightest Stars for the Construction of Mechanical Clocks, about 1550. Available at http://www.enotes.com/topic/Taqi_al-Din_Muhammad_ibn_Ma'ruf

Timebooth (2011) Ludwig Oechlin. http://www.timebooth.com/watchmakers/38/. Accessed 12 Dec 2011

Volker Vyskocil (2012) http://www.clockwatch.de/

Wikipedia (2001a) Galileo Galilei. http://en.wikipedia.org/wiki/Galileo_Galilei. Accessed 12 Dec 2011

Wikipedia (2001b) John Harrison. http://en.wikipedia.org/wiki/John_Harrison. Accessed 12 Dec 2011

Wikipedia (2001c) John Harrison. http://en.wikipedia.org/wiki/John_Harrison#The_first_three_marine_timekeepers. Accessed 12 Dec 2011

Wikipedia (2002a) Robert Hooke. http://en.wikipedia.org/wiki/Robert_Hooke. Accessed 12 Dec 2011

Wikipedia (2002b) Christiaan Huygens. http://en.wikipedia.org/wiki/Christiaan_Huygens. Accessed 12 Dec 2011

Wikipedia (2004a) Piazza dei Miracoli. http://en.wikipedia.org/wiki/Pisa_Cathedral#Duomo. Accessed 12 Dec 2011

Wikipedia (2004b) Verge escapement. http://en.wikipedia.org/wiki/Verge_escapement. Accessed 12 Dec 2011

Wikipedia (2004c) Anchor escapement. http://en.wikipedia.org/wiki/Anchor_escapement. Accessed 12 Dec 2011

Wikipedia (2004d) George Graham. http://en.wikipedia.org/wiki/George_Graham_(clockmaker). Accessed 12 Dec 2011

Wikipedia (2004e) Grasshopper Escapement. http://en.wikipedia.org/wiki/Grasshopper_escapement. Accessed 12 Dec 2011

Wikipedia (2004f) Escapement. http://en.wikipedia.org/wiki/Escapement#Detent_escapement. Accessed 12 Dec 2011

Wikipedia (2005) Thomas Earnshaw. http://en.wikipedia.org/wiki/Thomas_Earnshaw. Accessed 12 Dec 2011

Wikipedia (2006) George Daniels. http://en.wikipedia.org/wiki/George_Daniels_(watchmaker). Accessed 12 Dec 2011

Wikipedia (2008) Corpus Clock. http://en.wikipedia.org/wiki/Corpus_Clock. Accessed 12 Dec 2011

Wikipedia (2009) Pierre Le Roy. http://en.wikipedia.org/wiki/Pierre_Le_Roy. Accessed 12 Dec 2011

Chapter 3
The Mechanics of the Swiss Lever Escapement

As presented in the previous chapter, there have been many different kinds of escapements developed during the past six centuries. Though, at least 98% of the commercial mechanical watches today use the Swiss lever escapement. In this chapter, the Swiss Lever Escapement is studied in detail. Its working principle is illustrated and its dynamical model is derived. Experimental validation is also briefly discussed.

3.1 Introduction to the Swiss Lever Escapement

Generally, a mechanical watch is made of five parts as shown in Fig. 3.1. Among them the *escapement* is the most important, because it regulates the running of *gear train* to maintain a constant speed for *display* output. The energy lost during its oscillation is made up by the *mainspring* through *winding*, manual or automatic.

Figure 3.2 shows the Swiss lever escapement. It can be considered as the brain of the watch movement because it distributes the energy (the *distributing unit*, which includes an escape wheel and a pallet fork) and regulates the motion (the *regulating unit*, which includes a balance wheel and a hairspring); thus, controls the timekeeping accuracy. The name "escapement" arises from the fact that a small amount of energy is allowed to "escape" step by step.

In general, the movements of the escapement include the movements of the escape wheel, the pallet fork and the balance wheel. Both the balance wheel and the pallet fork swing while the escape wheel moves clockwise in an intermittent manner. For the majority of an oscillation cycle, the motions are rather "regular". However, in less than 10% of a cycle, the torque from the gear train is exerted to the escapement and shocks occur, which will change the motion. As shown in Fig. 3.3, a number of shocks occur in sequence in a half cycle (the other half is symmetric). Though, there are only three peaks in the sound wave as some shocks

R. Du and L. Xie, *The Mechanics of Mechanical Watches and Clocks*,
History of Mechanism and Machine Science 21, DOI: 10.1007/978-3-642-29308-5_3,

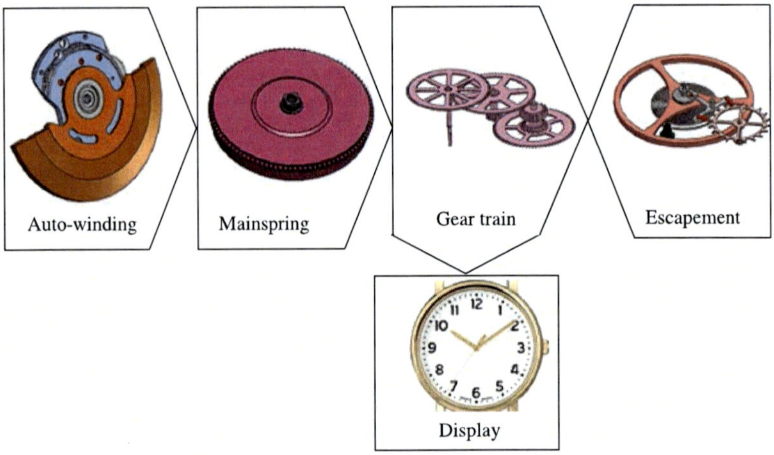

Fig. 3.1 Main components of mechanical watch movement

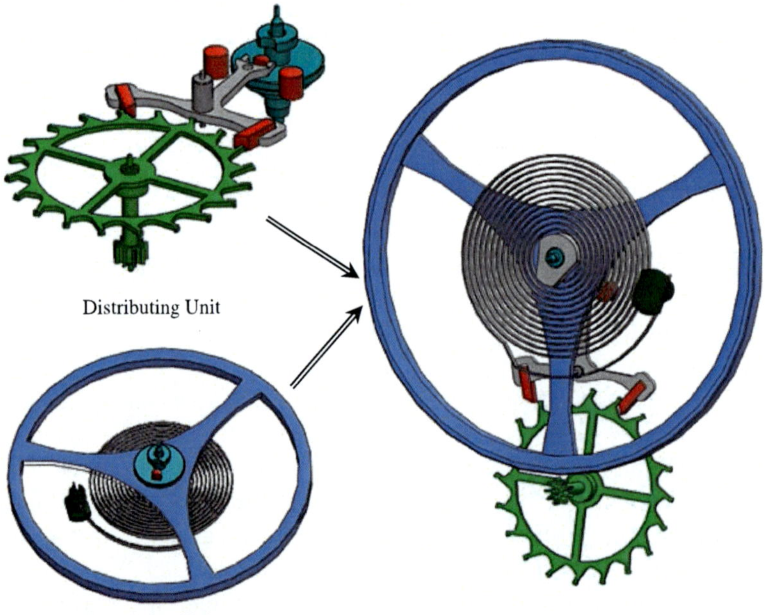

Fig. 3.2 The Swiss lever escapement

are very close to each other. Furthermore, we hear only one tick sound, which corresponds to the vibration of the escapement (the shock frequency is too low to hear by human ears). In the subsequent sections, we will derive the model of the Swiss lever escapement that describes the movements and the shocks.

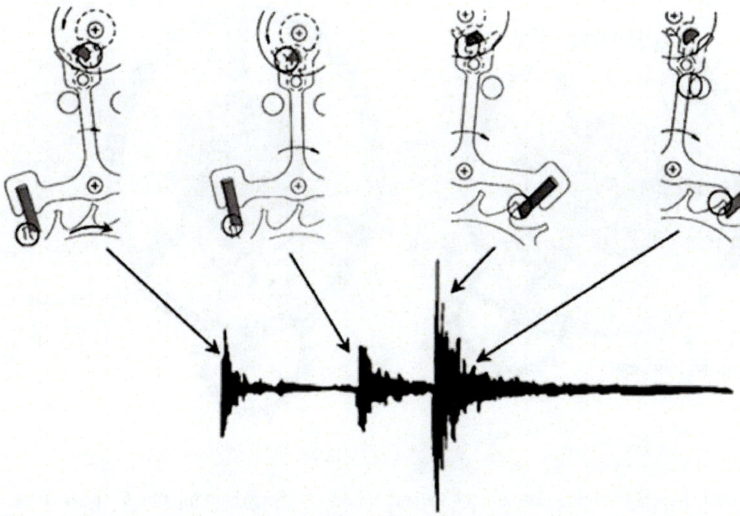

Fig. 3.3 The shocks in a half cycle of the Swiss lever escapement

3.2 The Motion of the Swiss Lever Escapement

Let us examine the motion of the Swiss lever escapement now. As detailed in Fig. 3.4, the escapement consists of five components: a balance wheel (on which there is an impulse pin), a hairspring, two banking pins (the entry banking pin and the exit banking pin), a pallet fork (on which there is an entry pallet jewel and an exit pallet jewel) and an escape wheel. Note that the outer end of the hairspring is fixed while the inner end is mounted onto the shaft of the balance wheel. The balance wheel periodically oscillates under the driving force from the escape wheel through the pallet fork as well as the restoring force of the hairspring. It is the impulse pin that sends and receives impulses from the pallet fork to the balance wheel. The two banking pins, on the other hand, limit the rotation of the pallet fork. The balance wheel swings at a specific speed according to the frequency of the hairspring. The calculation of this frequency is as follows, which is well known to the industry (Reymondin et al. 1999):

$$\mathrm{bph} = \frac{3600}{\pi}\sqrt{\frac{K}{J}} \tag{3.1}$$

where J is the moment of inertia of the balance wheel, K is the spring constant of the hairspring and bph stands for beat per hour.

The movements of the escapement include the angular displacement of the escape wheel, θ_e; the angular displacement of the pallet fork, θ_p; and the angular displacement of the balance wheel, θ_b. These movements are rather complex but can be approximated based on geometric analysis.

Figure 3.5 illustrates the operation of the escapement in first half of a cycle (the second half is symmetric in reverse order), in which there are a total of five

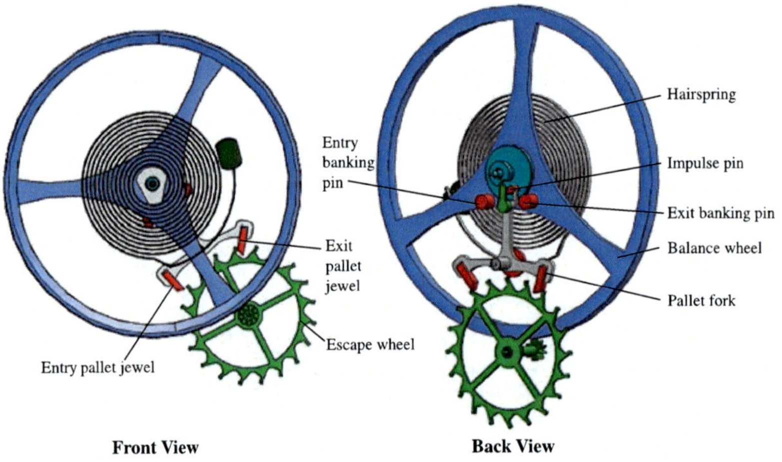

1 – hairspring, 2 – impulse pin, 3 – balance wheel, 4 – entry banking pin, 5 – pallet fork, 6 – exit pallet jewel, 7 – escape wheel, 8 – entry pallet jewel, 9 – exit banking pin

Fig. 3.4 Illustration of the Swiss lever escapement

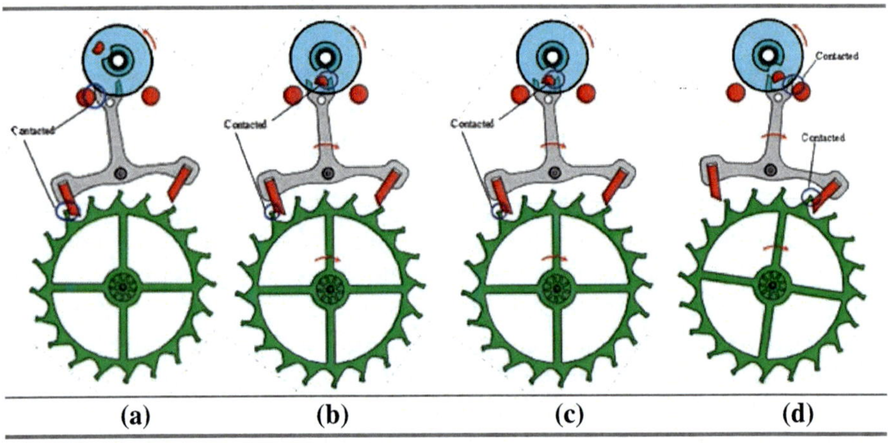

Fig. 3.5 The operation of the Swiss lever escapement in a half cycle

shocks but making four phases: (a) the lock phase, (b) the unlocking phase, (c) the impulse phase and (d) the safety phase. Note that in the figure, the shocks (which may transit from one phase to the subsequent phase) are marked by the *circles* and the directions of the rotation/swing are marked by *arrows*. Also, the time scale is not in proportion. The four phases are described as follows:

(a) The lock phase

Figure 3.6 shows the lock phase and its characteristics. A tooth of the escape wheel is locked on the entry pallet. In addition, the pallet fork is in contact with the entry banking pin. Therefore, both the escape wheel and the pallet fork are locked

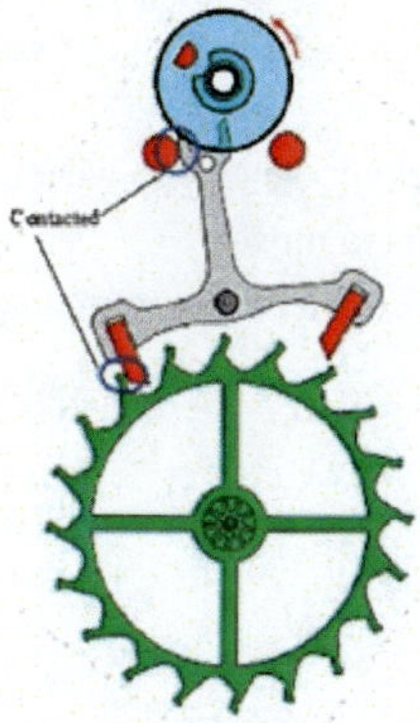

- The pallet fork is in contact with the entry blanking pin;
- A tooth of the escape wheel is in contact with the entry pallet;
- The escape wheel and the pallet fork are locked
- The balance wheel swings counter clockwise

Fig. 3.6 The main characteristics of the lock phase

not moving. Though, under the inertia force the balance wheel moves counter clockwise along the so-called supplementary arc and the velocity is increasing.

(b) The unlocking phase

Figure 3.7 shows the unlocking phase and its main characteristics. It starts from the first shock when the impulse pin on the balance wheel enters and impacts the pallet fork. This pushes the pallet fork swing clockwise to unlock the escape wheel. It allows the escape wheel turn clockwise and contacts the entry pallet.

Note that in the beginning of this phase, the pallet fork swings clockwise while it is in contact with the escape wheel; this causes the escape wheel recoil a small angle (about 0.25°). After the escape wheel disengages from the pallet fork, both the pallet fork and the escape wheel are rotating clockwise. Figure 3.8 shows the relative position between the pallet fork and the escape wheel during the unlocking phase. Note that θ_p and θ_e are negative. The key parameters include: R_e, R_8, β_1 and d_1. Figure 3.9 shows the geometric relationship of these parameters.

Let $\beta_1 = \theta_p - \beta_p$, where $\beta_p > 0$ is constant, then according to the geometric relationship in Fig. 3.9, the following relationship holds:

$$\frac{d_1 - R_e \cos\theta_e - R_8 \sin\beta_1}{R_8 \cos\beta_1 - R_e \sin\theta_e} = \frac{\cos\beta_1}{\sin\beta_1} \tag{3.2}$$

Solving this equation gives the angular displacement of the escape wheel:

$$\theta_e = \beta_1 - \sin^{-1}\frac{d_1 \sin\beta_1 + R_8}{R_e} \tag{3.3}$$

Figure 3.10 shows the relationship between the angular displacement of the pallet fork θ_p and the angular displacement of the escape wheel θ_e.

Furthermore, from the geometric relation in Fig. 3.8, the following equations can be obtained

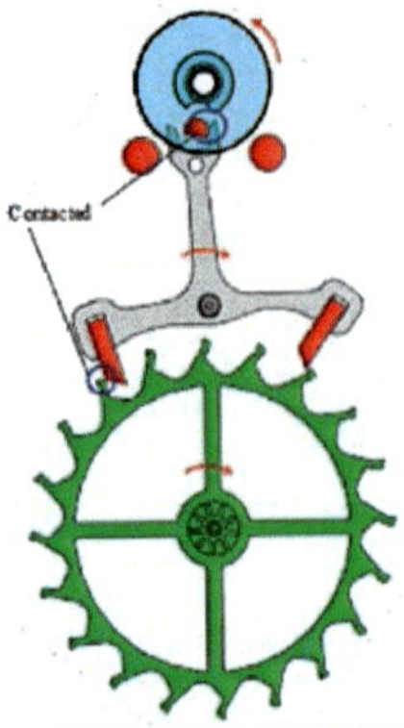

- The impulse pin on the balance wheel contacts the pallet fork, giving the first shock;
- The pallet fork moves clockwise unlocking the escape wheel;
- The escape wheel is unlocked and starts to move;
- The balance wheel continues to swing counter-clockwise.

Fig. 3.7 The characteristics of the unlocking phase

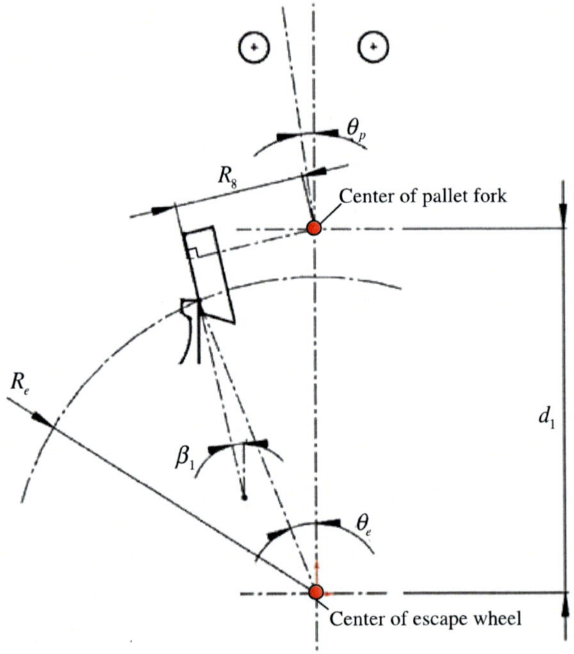

Fig. 3.8 Schematics of the escape wheel and the pallet fork in the unlocking phase

$$R_e \sin\theta_e = -R_8 \cos\beta_1 - R_6 \sin\beta_1 \tag{3.4}$$

$$R_e \cos\theta_e = d_1 + R_8 \sin\beta_1 - R_6 \cos\beta_1 \tag{3.5}$$

$$R_6^2 - 2R_6 d_1 \cos\beta + d_1^2 + R_8^2 - R_e^2 + 2d_1 R_8 \sin\beta_1 = 0 \tag{3.6}$$

Thus, the parameters R_5, R_6 and R_7 are

$$R_5 = \sqrt{R_e^2 - R_8^2 - d_1^2 \sin^2\beta_1 - 2d_1 R_8 \sin\beta_1} \tag{3.7}$$

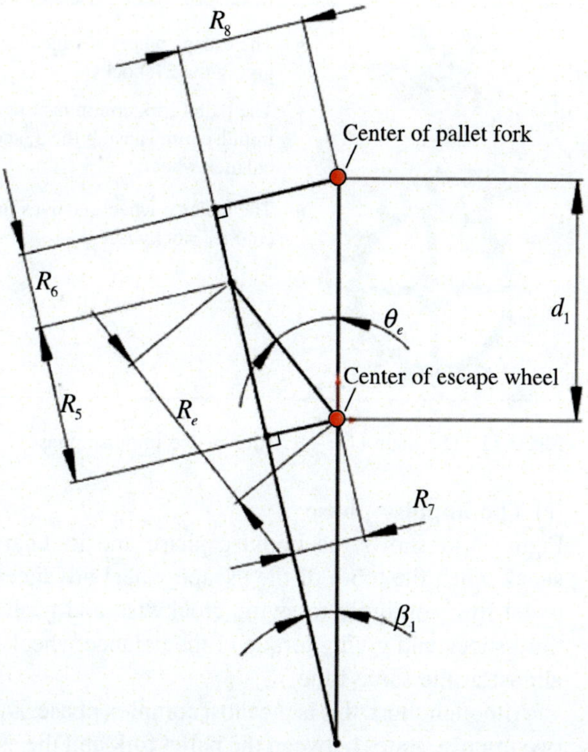

Fig. 3.9 Key dimensions, R_e, R_8 and d_1, during the unlocking phase

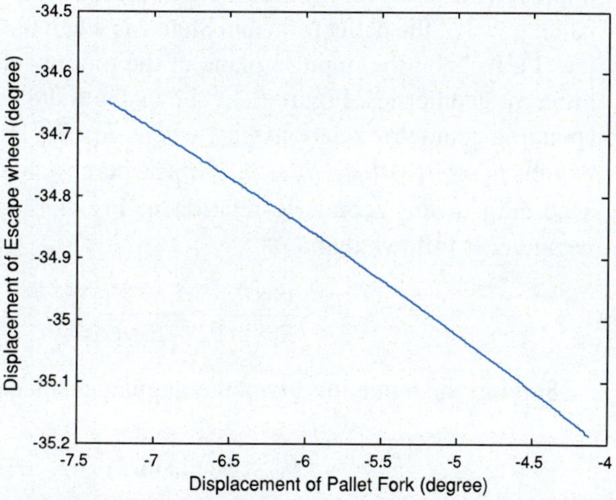

Fig. 3.10 The relative motion between the pallet fork and the escape wheel during the unlocking phase

$$R_6 = d_1 \cos \beta_1 - \sqrt{R_e^2 - R_8^2 - d_1^2 \sin^2 \beta_1 - 2d_1 R_8 \sin \beta_1} \tag{3.8}$$

$$R_7 = R_8 + d_1 \sin \beta_1 \tag{3.9}$$

These parameters determine the relative position between the escape wheel and the pallet fork.

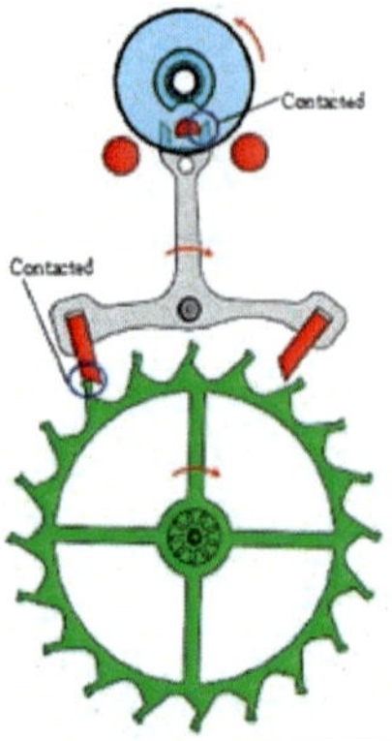

- The escape wheel advances and its tooth impacts the entry pallet giving the 2nd shocks;
- The pallet fork continues to swing clockwise and impacts the impulse pin, causing the 3rd shock and giving energy to the balance wheel;
- The balance wheel receives the energy and continues to swing counter clockwise.

Fig. 3.11 The main characteristics of the impulse phase

(c) The impulse phase

Figure 3.11 shows the impulse phase and its characteristics. It starts the second shock when the tooth of the escape wheel advances to impact the pallet fork. The pallet fork continues to swing clockwise and pushes the impulse pin, causing the third shock and giving torque to the balance wheel. Note that the two shocks occur almost at the same time.

Although short, this is the most complex phase. During the second shock, there are two impulse states between the pallet fork and the escape wheel: State 1 is when the impulse beak of the tooth of the escape wheel travels on the impulse plane of the entry pallet jewel of the pallet fork, and State 2 is when the impulse beak of the entry pallet jewel travels on the impulse plane of the tooth of the escape wheel. They result in different geometries. Figure 3.12 shows the State 1 and Fig. 3.13 shows the corresponding geometric relationships, where, R_e, R_{10} and d_1 are design parameters.

Let $\beta_2 = \beta_1 - \beta_j = \theta_p - \beta_p - \beta_j$, where $\beta_j > 0$ is constant parameter, then according to the geometric relation in Fig. 3.13, and noting that θ_p and θ_e are negative, it follows that:

$$\frac{R_e \cos\theta_e - d_1 - R_{10}\sin\beta_2}{R_e \sin\theta_e + R_{10}\cos\beta_2} = \frac{\cos\beta_2}{\sin\beta_2} \tag{3.10}$$

Solving this equation gives the angular displacement of the escape wheel:

$$\theta_e = \beta_2 - \sin^{-1}\frac{d_1 \sin\beta_2 + R_{10}}{R_e} \tag{3.11}$$

Figure 3.14 shows the relationship between the angular displacement of the pallet fork and the angular displacement of the escape wheel.

Furthermore, the following equations can be obtained:

$$R_e \sin\theta_e = R_2 \sin\beta_2 - R_{10}\cos\beta_2 \tag{3.12}$$

$$R_e \cos\theta_e = d_1 + R_{10}\sin\beta_2 + R_2\cos\beta_2 \tag{3.13}$$

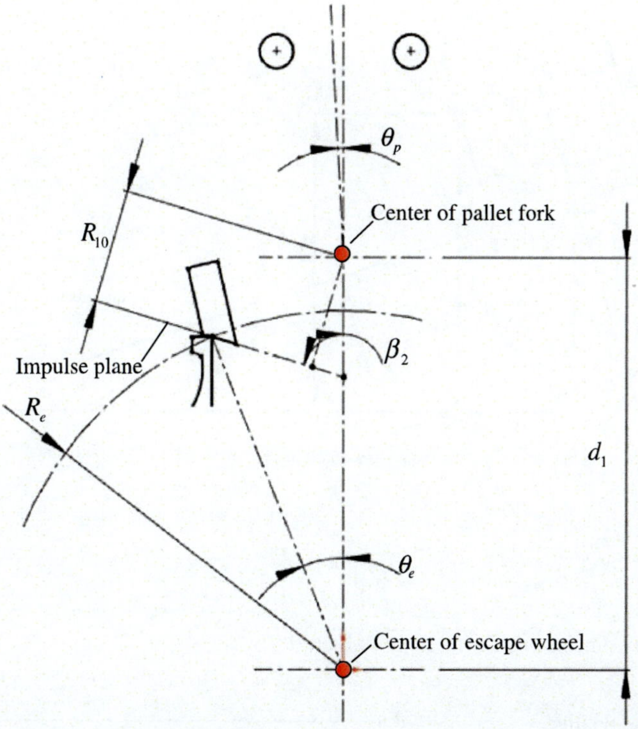

Fig. 3.12 Schematics of the escape wheel and the pallet fork during impulse State 1

$$R_2^2 + 2R_2d_1\cos\beta_2 + R_{10}^2 + d_1^2 - R_e^2 + 2R_{10}d_1\sin\beta_2 = 0 \tag{3.14}$$

Therefore, the location parameters, R_1, R_2 and R_9 are as follows:

$$R_2 = -d_1\cos\beta_2 + \sqrt{R_e^2 - R_{10}^2 - d_1^2\sin^2\beta_2 - 2d_1R_{10}\sin\beta_2} \tag{3.15}$$

$$R_1 = \sqrt{R_e^2 - R_{10}^2 - d_1^2\sin^2\beta_2 - 2d_1R_{10}\sin\beta_2} \tag{3.16}$$

$$R_9 = -R_{10} - d_1\sin\beta_2 \tag{3.17}$$

Figure 3.15 shows the relative position of the pallet fork and the escape wheel in State 2, and Fig. 3.16 illustrates the corresponding critical dimensions, where R_p, R_{11} and d_1 are design parameters.

Let $\beta_3 = \theta_p + \beta_k$ and $\alpha = \theta_e + \alpha_k$, where $\beta_k > 0$ and $\alpha_k > 0$ are constants, then it can be shown that:

$$\frac{d_1 - R_p\cos\beta_3 - R_{11}\cos\alpha}{-R_p\sin\beta_3 - R_{11}\sin\alpha} = -\frac{\sin\alpha}{\cos\alpha} \tag{3.18}$$

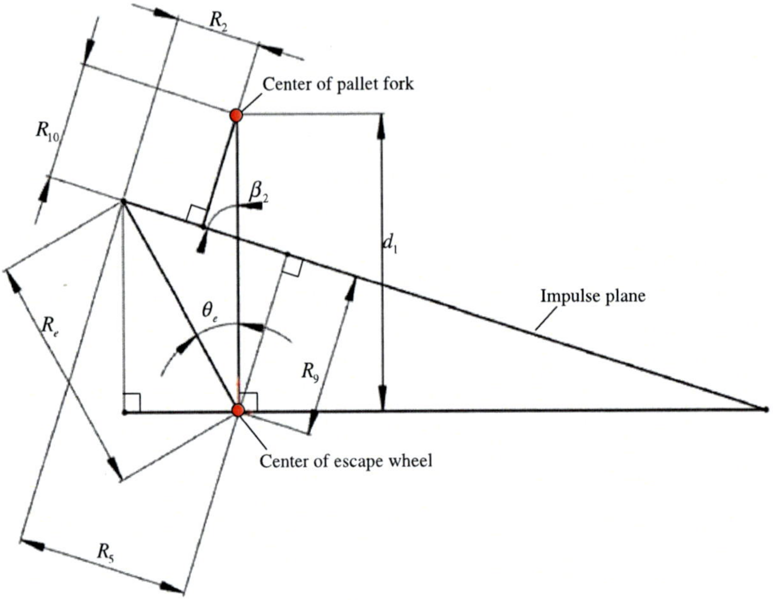

Fig. 3.13 Key dimensions during impulse State 1

Fig. 3.14 The relative motion between the pallet fork and the escape wheel

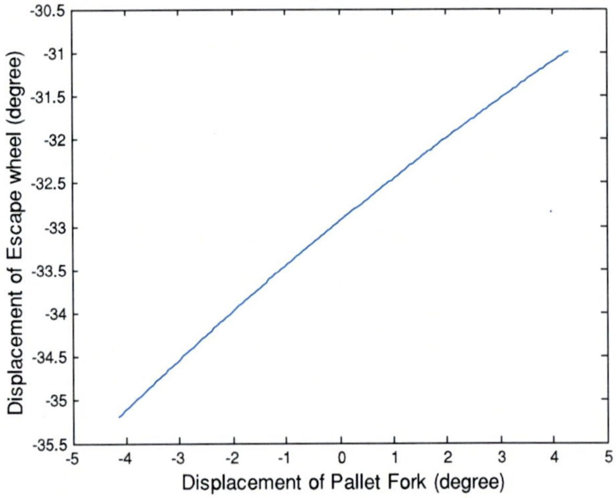

Solving the above equation gives the angular displacement of the escape wheel:

$$\alpha = -\tan^{-1}\left(\frac{R_p \cos\beta_3 - d_1}{R_p \sin\beta_3}\right) - \sin^{-1}\left(\frac{R_{11}}{\sqrt{(R_p \cos\beta_3 - d_1)^2 + (R_p \sin\beta_3)^2}}\right) \tag{3.19}$$

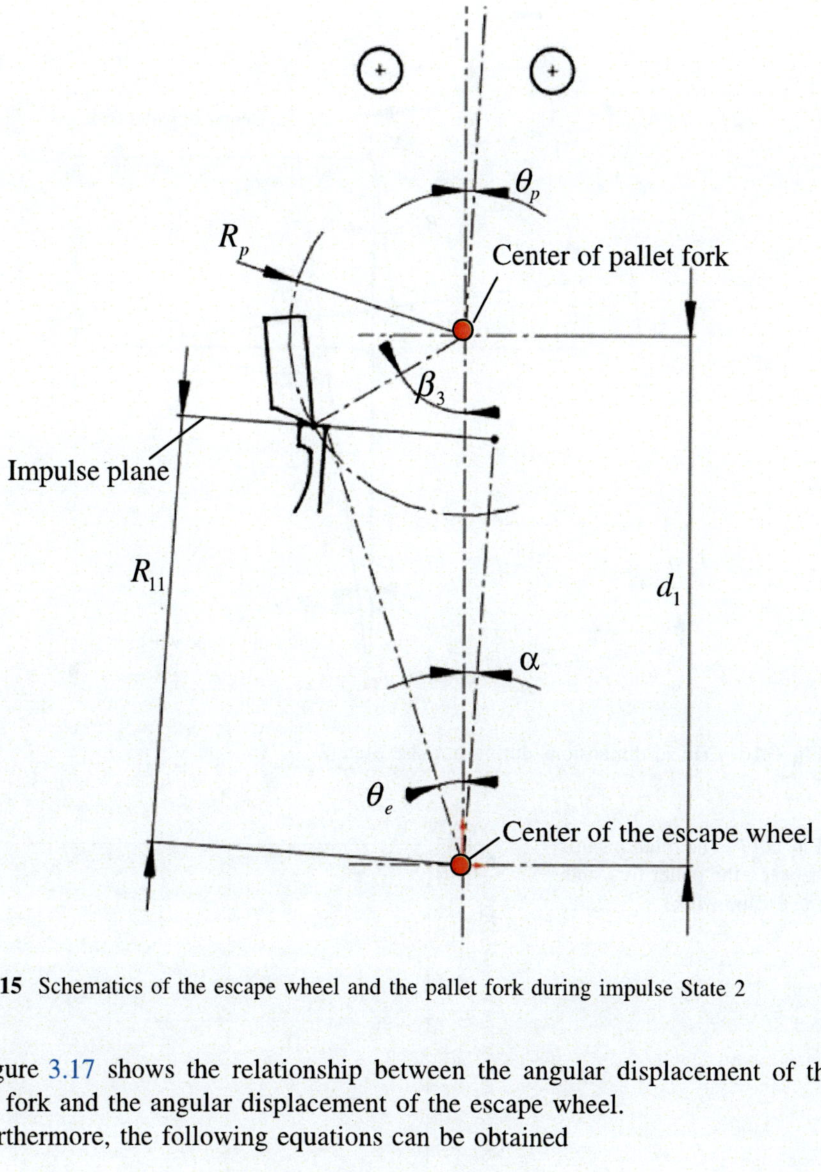

Fig. 3.15 Schematics of the escape wheel and the pallet fork during impulse State 2

Figure 3.17 shows the relationship between the angular displacement of the pallet fork and the angular displacement of the escape wheel.

Furthermore, the following equations can be obtained

$$R_p \sin \beta_3 = R_{13} \cos \alpha - R_{11} \sin \alpha \tag{3.20}$$

$$R_p \cos \beta_3 = d_1 - R_{13} \sin \alpha - R_{11} \cos \alpha \tag{3.21}$$

$$R_{13}^2 - 2R_{13}d_1 \sin \alpha + R_{11}^2 + d_1^2 - R_p^2 - 2R_{11}d_1 \cos \alpha = 0 \tag{3.22}$$

Therefore, the location parameters R_{12}, R_{13} and R_{14} are as follows:

$$R_{13} = d_1 \sin \alpha + \sqrt{R_p^2 - R_{11}^2 - d_1^2 \cos^2 \alpha + 2d_1 R_{11} \cos \alpha} \tag{3.23}$$

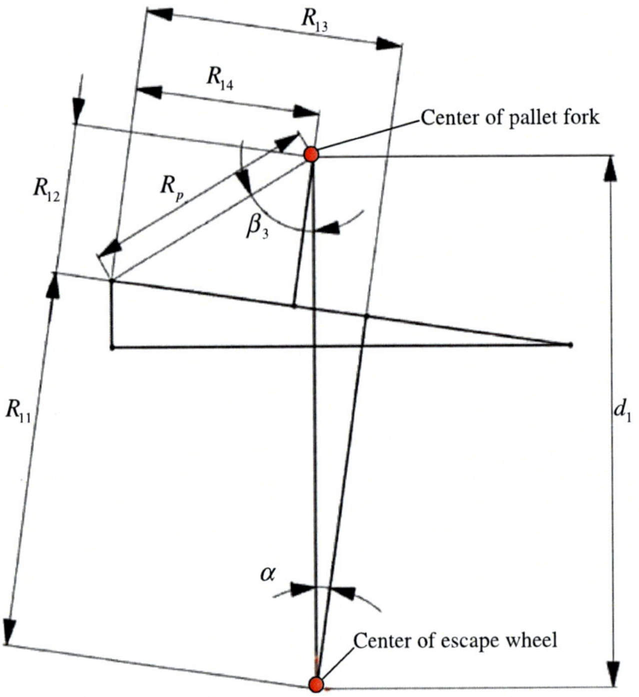

Fig. 3.16 Critical dimensions during impulse State 2

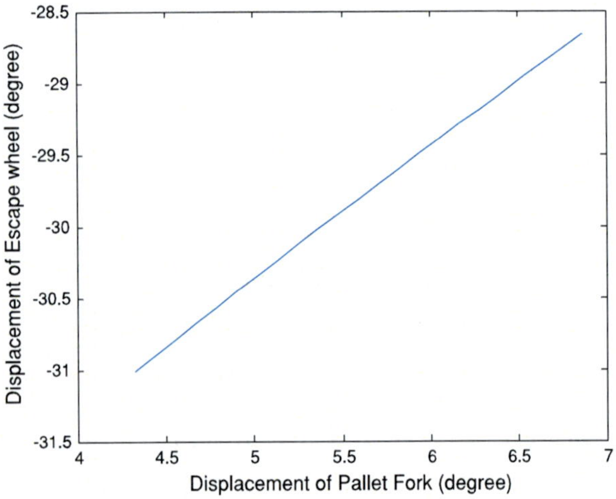

Fig. 3.17 The relationship between the pallet fork and the escape wheel

$$R_{14} = \sqrt{R_p^2 - R_{11}^2 - d_1^2 \cos^2 \alpha + 2d_1 R_{11} \cos \alpha} \tag{3.24}$$

$$R_{12} = d_1 \cos \alpha - R_{11} \tag{3.25}$$

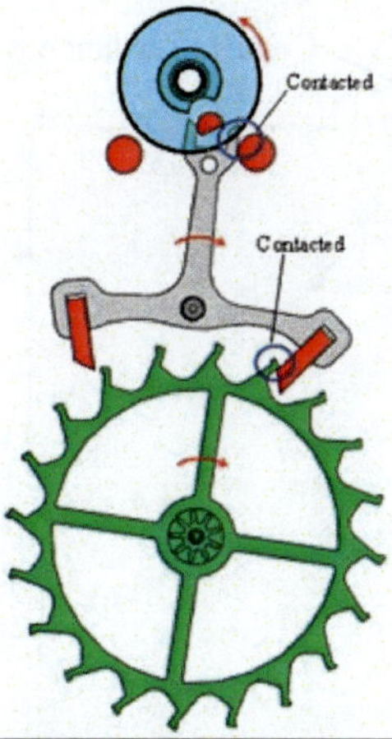

- The escape wheel continues moving forward and contacts the exit pallet, causing the 4th shock;
- The pallet fork contacts the blanking pin, causing the 5th shock;
- Both the escape wheel and the pallet fork are locked
- The balance wheel continues to swing counter clockwise until it reaches its extreme location, and the start to swing back

Fig. 3.18 The characteristics of the safety phase

(d) The safety phase

Figure 3.18 shows the characteristics of the safety phase. It starts with the fourth shock when the escape wheel impacts the exit pallet and ends when the pallet fork hits the blanking pin, which gives the fifth shock. At this time, both the pallet fork and the escape wheel are locked. Though, the balance wheel continues to swing following the so-called the supplementary arc, but the speed is descending. When the inertial force is exhausted, the balance wheel finishes the half-period and returns in a counter-clockwise direction following the same phases in reversed order. This is the lock phase at the beginning of the second half of the cycle.

It should be mentioned that during the entire half cycle, the relative position between the balance wheel and the pallet fork does not change during the entire cycle. As shown in Fig. 3.19, R_b is the distance between the center of the balance wheel and the center of the impulse pin, and d_2 is the distance between the center of balance wheel and the center of the pallet fork. These two parameters are design parameters.

Based on Fig. 3.19, it can be shown that the following relationship holds:

$$\frac{d_2 - R_b \cos\theta_b}{-R_b \sin\theta_b} = \frac{\cos\theta_p}{\sin\theta_p} \tag{3.26}$$

This gives

$$\theta_b = \theta_p - \sin^{-1}\left(\frac{d_2}{R_b}\sin\theta_p\right) \tag{3.27}$$

Figure 3.20 shows the relationship between the angular displacement of the pallet fork and the angular displacement of the balance wheel.

In summary, based on the geometric relations above and assume the movement of balance wheel is sinusoidal, we can then find the movements of the pallet fork and the escape wheel. Figure 3.21 shows the angular displacements of the balance wheel, the pallet fork and the escape wheel in a cycle. From the figure, it is seen

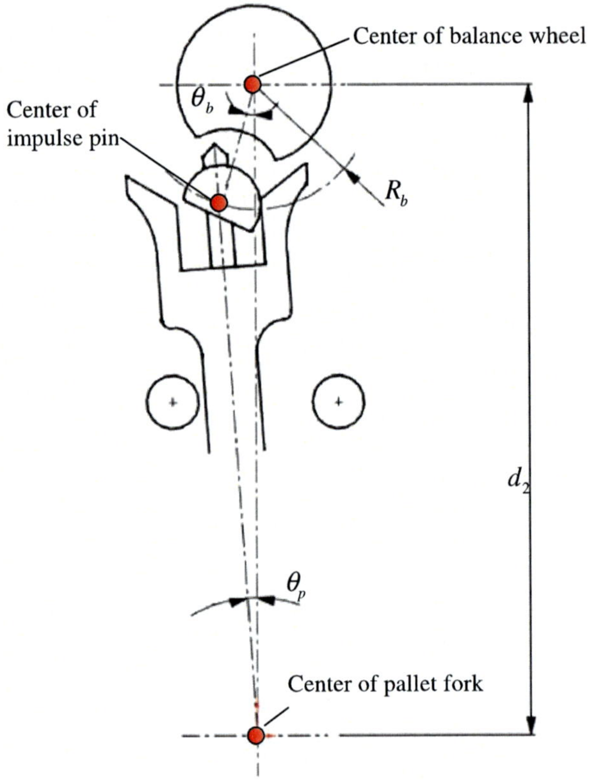

Fig. 3.19 Schematics of the balance wheel and the pallet fork

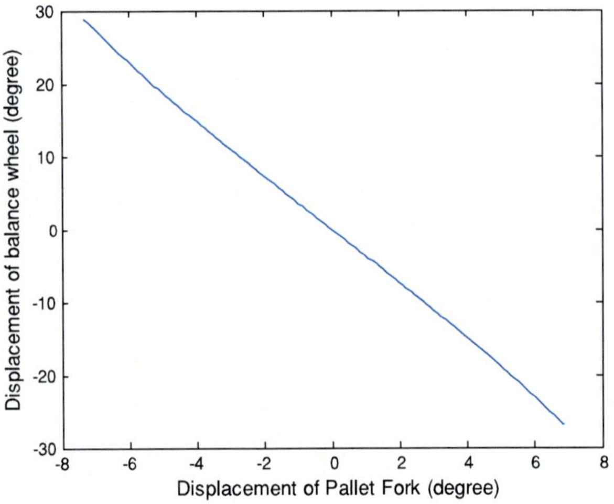

Fig. 3.20 The relationship between the pallet fork and the balance wheel

that the pallet fork moves in a trapezoidal form, the two levels correspond to the two locking positions. The escape wheel moves step by step, the step corresponds to the locking and the small dip before the raise corresponds to the recoil.

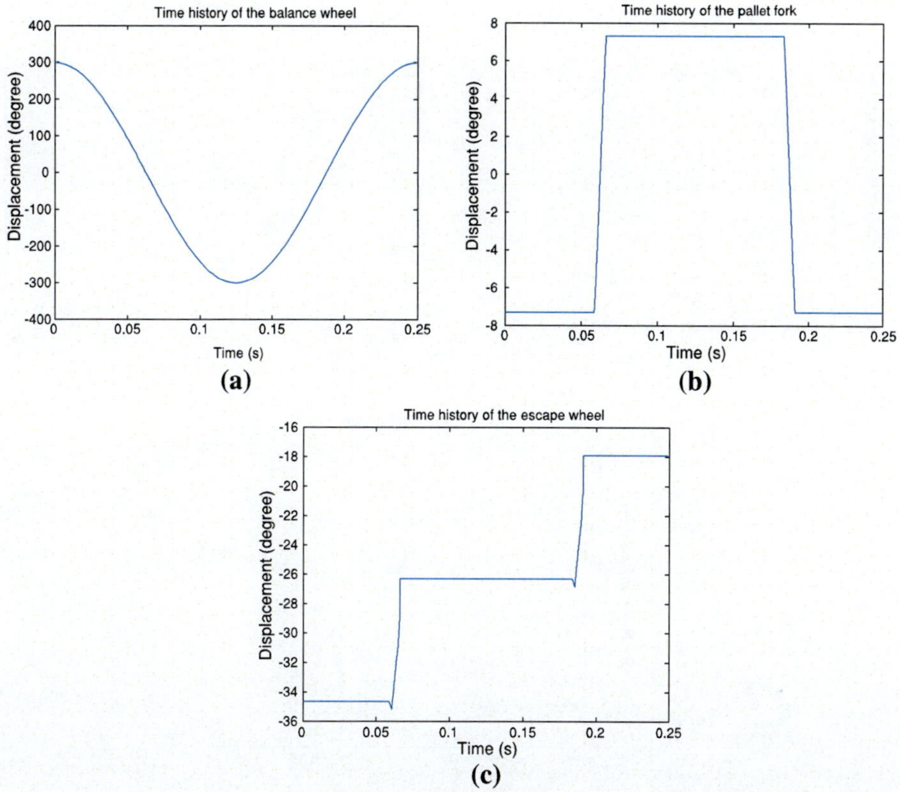

Fig. 3.21 The movement of the Swiss lever escapement in a cycle. **a** The angular displacement of the balance wheel. **b** The angular displacement of the pallet fork. **c** The angular displacement of the escape wheel

The most important period is the unlocking phase and the impulse phase. Figure 3.22 zooms into this period. Note that $t_1, t_2, \ldots, t_5$ are the time instances of the five shocks in the four phases. It should be pointed out that the shocks all occur in a short period of time less than 10% of a cycle. Yet, they complete the energy transfer from the escape wheel to the balance wheel, and are very important to the Swiss lever escapement.

Figure 3.23 summarizes the timeline of movement events (note that it is not on scale). This also lays the foundation for the dynamic model presented in the next section.

3.3 Dynamic Modeling by Impulsive Differential Equations

In the previous section, the movement of the Swiss lever escapement is studied based on the geometry. It gives a good approximation. However, the escapement is not a simple rigid body and its dynamics must be taken into consideration.

According to the literature, one of the earliest studies on the dynamics of mechanical clock was done by Kauderer (1958). He used a set of piece-wise linear

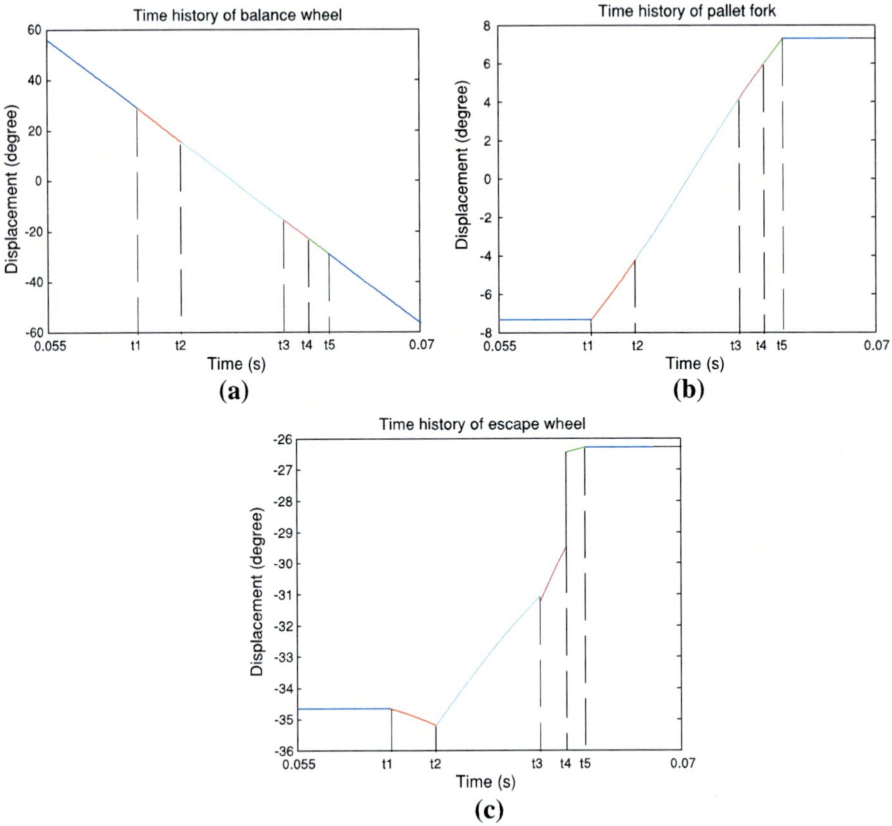

Fig. 3.22 The angular motion of the Swiss lever escapement in the unlocking and impulse phase. **a** The displacement of the balance wheel. **b** The displacement of the pallet fork. **c** The angular displacement of the escape wheel

differential equations to model a pendulum clock. In a recent study, Schwatz and Gran (2011) did the same using MATLAB®. However, as there are many more impulses in the Swiss lever escapement, the piece-wise linear differential equation becomes imperative.

To deal with the complex impact mechanics, in recent years, a number of methods have been developed, such as the differential inclusion (Manuel et al. 1993), and the constraint enforcement (Borri et al. 1990). We adopted the impulsive differential equation. This technique was firstly proposed in 1980s (Lakshmikantham et al. 1989). Briefly, the impulsive differential equation takes the following form:

$$\frac{dx}{dt} = f(t, x), \text{ if } \phi(x) \neq 0 \tag{3.28}$$

$$\Delta x = \rho(x), \text{ if } \phi(x) = 0 \tag{3.29}$$

$$x(t_0^+) = x_0 \tag{3.30}$$

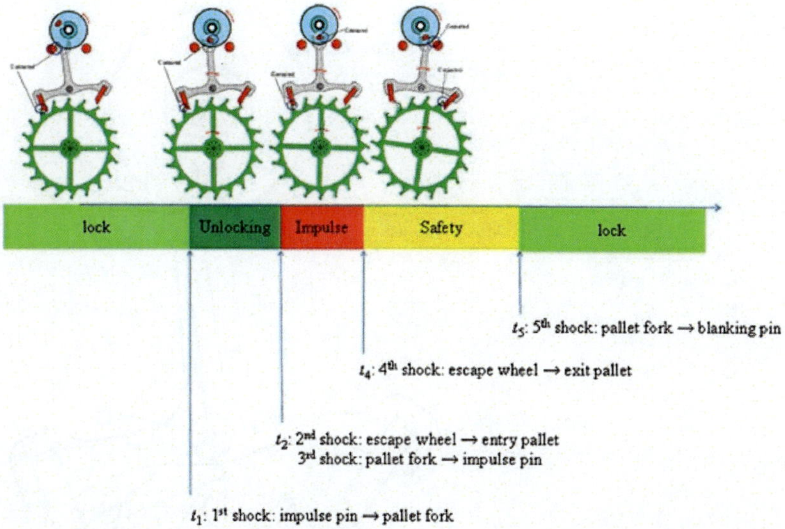

Fig. 3.23 The timeline of the escapement movement

where f: $\mathbb{R} \times \Omega \to \mathbb{R}^n$, $\{x|\phi(x) = 0\}$ is the state function of the system, ρ: $\mathbb{R} \times \Omega \to \Omega$ is the jump function and Ω is the set of the states of the system. Its difference to the ordinary differential equation lays on the jump function.

The impulsive differential equation can be applied to solve problems that involve many impacts. In Roup et al. (2001), it is used to study the Verge escapement. Solving the Swiss lever escapement problem is, however, much more complicated.

To derive the dynamic model of the Swiss lever escapement, following assumptions are made:

(a) There is no deformation in the escapement except for the hairspring.
(b) There is no friction between the axles and the bearings.
(c) The center of gravity of the balance wheel is on the axis of rotation.
(d) The relationship between the hairspring moment and the angle displacement of the balance wheel is linear.
(e) The weight of the hairspring can be ignored.

Figure 3.24 shows the more detailed definitions of the geometric parameters of the escapement. The key parameters include the following:

r_1—the contact radius of the escape wheel during impulse between the escape wheel and the pallet fork;
r_2—the contact radius of the pallet fork during impulse between the escape wheel and the pallet fork;
r_3—the contact radius of the pallet fork during the impulse between the pallet fork and the balance wheel;
r_4—the contact radius of the balance wheel during the impulse between the pallet fork and the balance wheel;

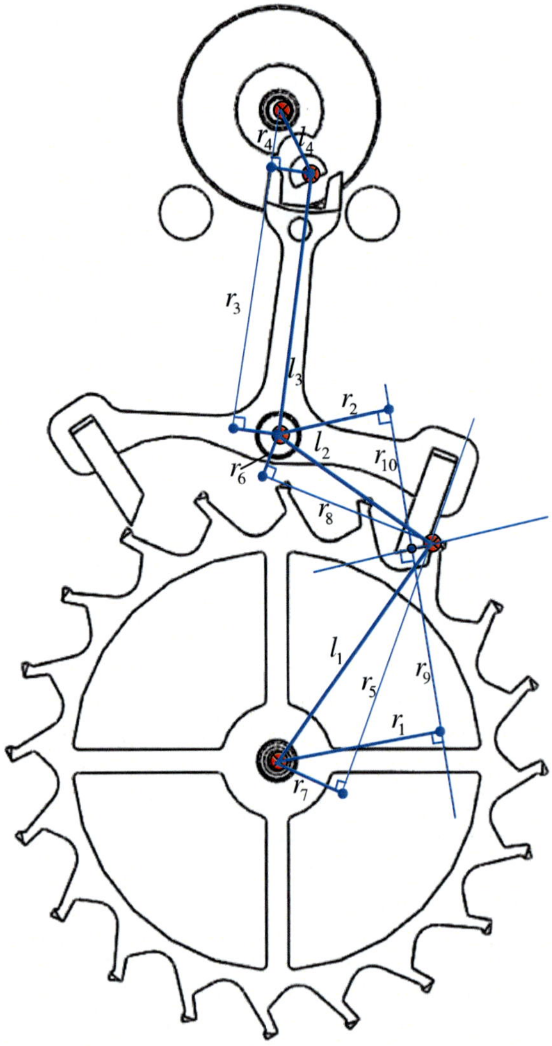

Fig. 3.24 Definitions of the key parameters in the model

r_5—the contact radius of the escape wheel during unlocking between the escape wheel and the pallet fork;
r_6—the contact radius of the pallet fork during unlocking between the escape wheel and the pallet fork;
r_7—the friction radius of the escape wheel during unlocking between the escape wheel and the pallet fork;
r_8—the friction radius of the pallet fork during unlocking between the escape wheel and the pallet fork;
r_9—the friction radius of escape wheel during impulse between the escape wheel and the pallet fork;
r_{10}—the friction radius of the pallet fork during impulse between the escape wheel and the pallet fork;

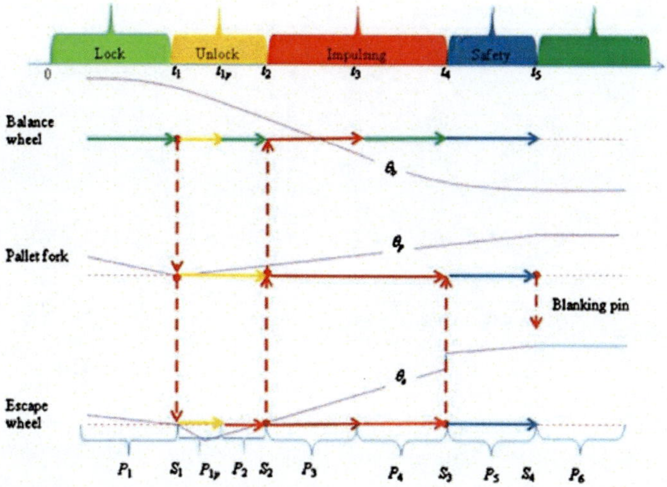

Fig. 3.25 Illustration of the 11 states

l_1—the distances from the pallet to the axle of the escape wheel;
l_2—the distances from the pallet to the axle of the pallet fork;
l_3—the distances from the impulse pin to the axle of the pallet fork;
l_4—the distances from the impulse pin to the axle of the balance wheel;
θ_b—the angular displacement of balance wheel;
θ_p—the angular displacement of pallet fork;
θ_e—the angular displacement of escape wheel;
$\dot{\theta}_b$—the angular velocity of balance wheel;
$\dot{\theta}_p$—the angular velocity of pallet fork;
$\dot{\theta}_e$—the angular velocity of escape wheel;
$\ddot{\theta}_b$—the angular acceleration of balance wheel;
$\ddot{\theta}_p$—the angular acceleration of pallet fork;
$\ddot{\theta}_e$—the angular acceleration of escape wheel;
θ_{b0}—the initial angular displacement;
$\theta_{b1,}\theta_{b5}$—thresholds defined by the guard pins;
J_b—the inertia of the balance wheel;
J_p—the inertia of the pallet fork;
J_e—the inertia of the escape wheel; and
τ—the torque exerted on the escape wheel.

Because the motion is rather symmetric, for convenience, define

$$\sigma = \begin{cases} +1, & \text{1st half-period} \\ -1, & \text{2nd half-period} \end{cases} \tag{3.31}$$

This will simplify the mathematical representation.

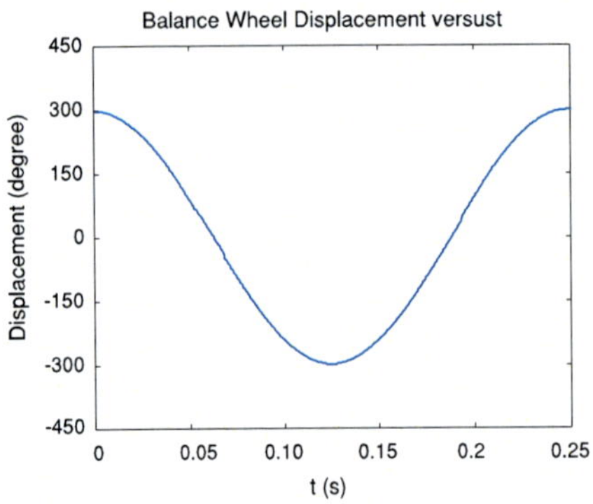

(a)

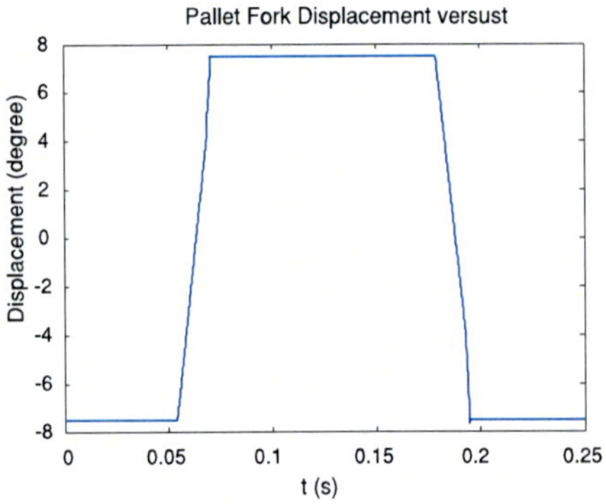

(b)

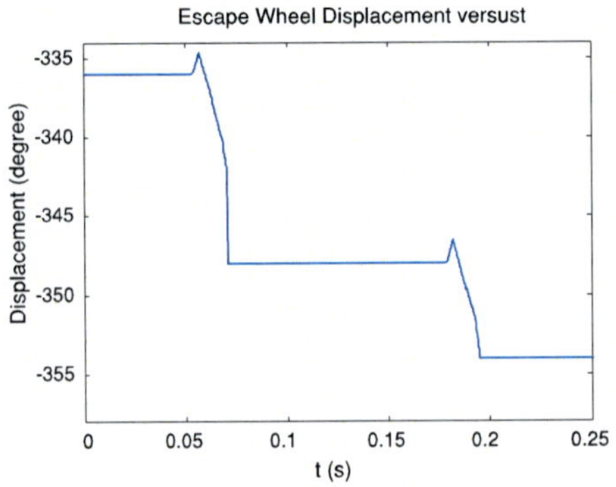

(c)

Fig. 3.26 The angular displacement in a cycle. **a** The balance wheel. **b** The pallet fork. **c** The escape wheel

As shown in the previous section, there are four phases containing five shocks in half a cycle. Together with the period between shocks, there are totally 11 states in half a cycle. Some of these states are time periods (which will be denoted as *P*) and the others are time instances (which will be denoted as *S*). The following are the models of each state.

(a) Lock phase

State 1 (P_1, $t = (0 \sim t_1)$): The balance wheel follows the ascending supplementary arc while the pallet fork and the escape wheel are locked. The movement of the balance wheel is governed by the equation:

$$J_b\ddot{\theta}_b + K\theta_b = 0 \tag{3.32}$$

The initial condition is $\theta_b = \theta_{b0}, \dot{\theta}_b = 0$, at $t = 0$. This step ends when the impulse pin reaches the pallet fork giving the first shock at the time t_1. At this time, the angle and velocity of the balance wheel are θ_{b1} and $\dot{\theta}_{b10}$, respectively. Note that t_1 is a function of θ_{b1} and the initial conditions.

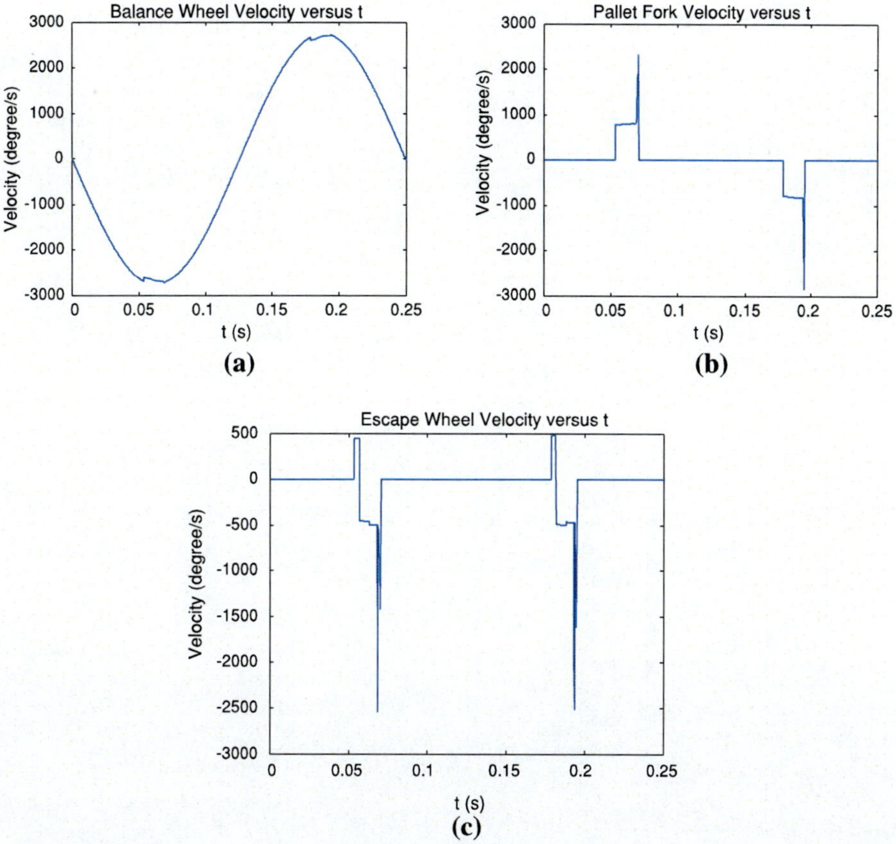

Fig. 3.27 The angular velocity in a cycle. **a** The balance wheel. **b** The pallet fork. **c** The escape wheel

(b) Unlocking phase

State 2 (S_1, $t = t_1$): At t_1, the first shock occurs. The collision is assumed to be inelastic and the resulting relationship among the balance wheel, the pallet fork and the escape wheel is as follows:

$$\frac{J_b}{r_4 r_6}\dot{\theta}_{b10} = \frac{J_b}{r_4 r_6}\dot{\theta}_{b11} - \frac{J_p}{r_3 r_6}\dot{\theta}_{p1} + \sigma\frac{J_e}{r_3 r_5}\dot{\theta}_{e1} \tag{3.33}$$

Because

$$\frac{\dot{\theta}_{b11}}{\dot{\theta}_{p1}} = -\frac{l_3}{l_4} \tag{3.34}$$

$$\frac{\dot{\theta}_{p1}}{\dot{\theta}_{e1}} = \sigma\frac{l_1}{l_2} \tag{3.35}$$

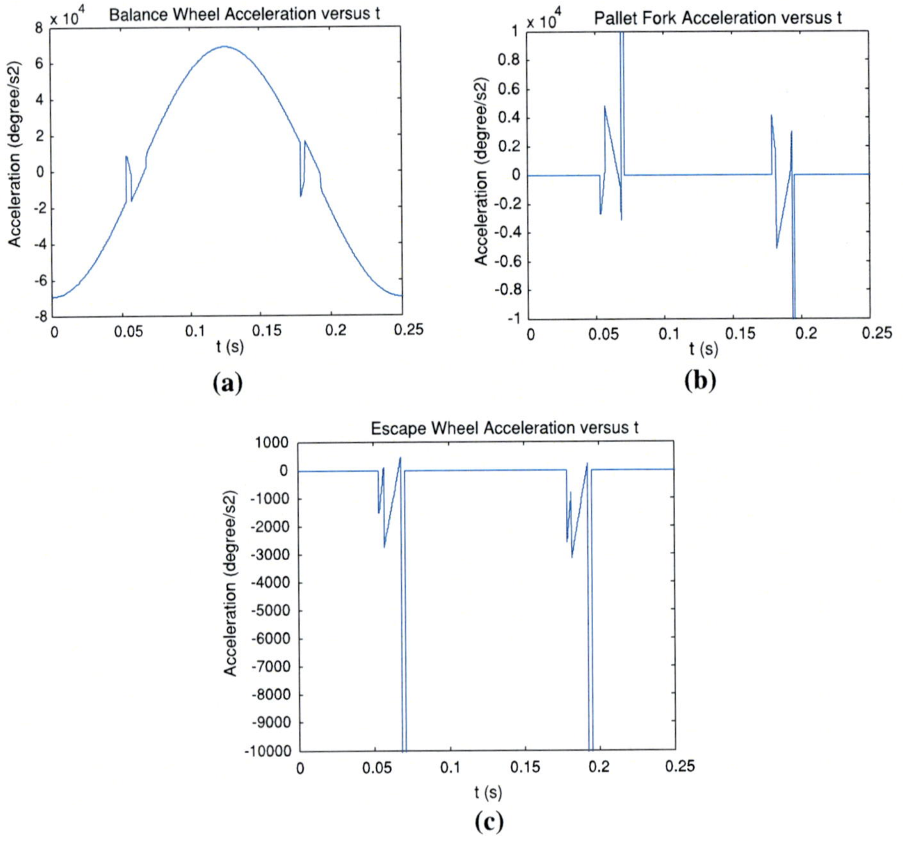

Fig. 3.28 The angular acceleration in a cycle. **a** The balance whell. **b** The pallet fork. **c** The escape wheel

It follows that

$$\begin{aligned}\frac{J_b}{r_4 r_6}\dot{\theta}_{b10} &= \left(\frac{J_b}{r_4 r_6} + \frac{J_p l_4}{r_3 r_6 l_3} + \frac{J_e l_2 l_4}{r_3 r_5 l_1 l_3}\right)\dot{\theta}_{b11} \\ &= -\left(\frac{J_b l_3}{r_4 r_6 l_4} + \frac{J_p}{r_3 r_6} + \frac{J_e l_2}{r_3 r_5 l_1}\right)\dot{\theta}_{p1} \\ &= \sigma\left(\frac{J_b l_1 l_3}{r_4 r_6 l_2 l_4} + \frac{J_p l_1}{r_3 r_6 l_2} + \frac{J_e}{r_3 r_5}\right)\dot{\theta}_{e1}\end{aligned} \tag{3.36}$$

Therefore,

$$\Delta\dot{\theta}_b = -\frac{J_p r_4 r_5 l_1 l_4 + J_e r_4 r_6 l_2 l_4}{J_b r_3 r_5 l_1 l_3 + J_p r_4 r_5 l_1 l_4 + J_e r_4 r_6 l_2 l_4}\dot{\theta}_{b10} \tag{3.37}$$

$$\Delta\dot{\theta}_p = -\frac{J_b r_3 r_5 l_1 l_4}{J_b r_3 r_5 l_1 l_3 + J_p r_4 r_5 l_1 l_4 + J_e r_4 r_6 l_2 l_4}\dot{\theta}_{b0} \tag{3.38}$$

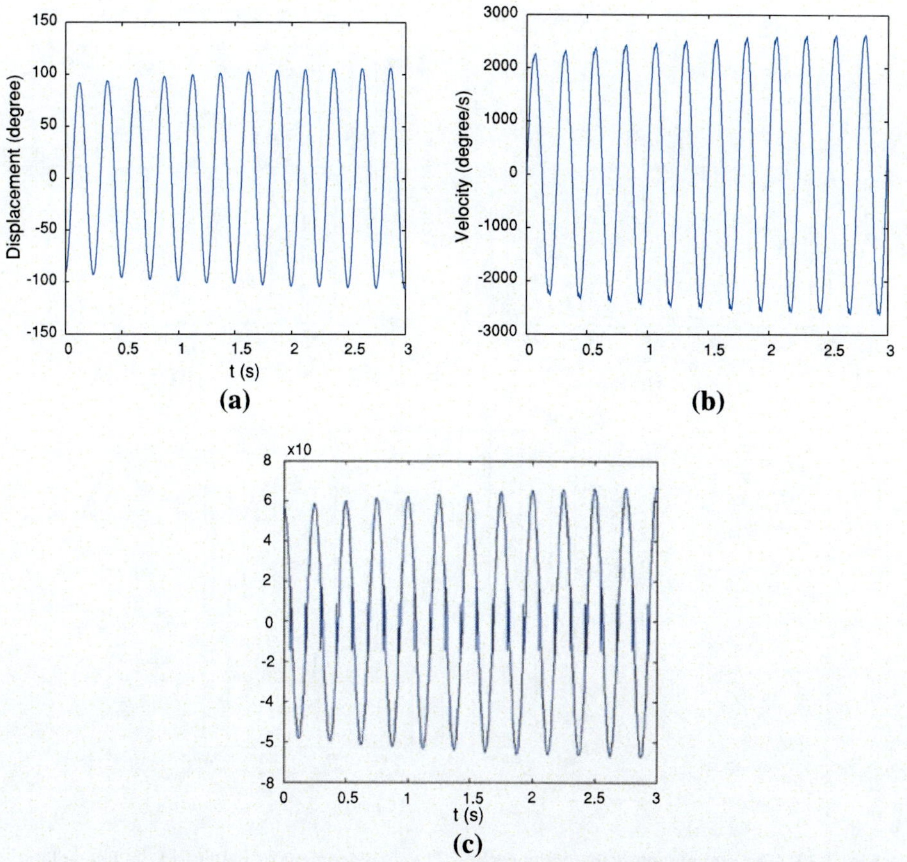

Fig. 3.29 The movement of the balance wheel. **a** The displacement of balance wheel. **b** The velocity of balance wheel. **c** The acceleration of balance wheel

$$\Delta\dot{\theta}_e = \sigma \frac{J_b r_3 r_5 l_2 l_4}{J_b r_3 r_5 l_1 l_3 + J_p r_4 r_5 l_1 l_4 + J_e r_4 r_6 l_2 l_4} \dot{\theta}_{b0} \tag{3.39}$$

where $\Delta\dot{\theta} \equiv \dot{\theta}_{11} - \dot{\theta}_{10}$, $\Delta\dot{\theta}_p \equiv \dot{\theta}_{p1}$ and $\Delta\dot{\theta}_e \equiv \dot{\theta}_{e1}$.

State 3 (P_{1p}, $t = (t_1 \sim t_{1p})$): After the first shock, the balance wheel pushes escape wheel via pallet fork. Hence, the resulting equation is

$$J_1 \ddot{\theta}_b + K\theta_b = -\sigma T_1 \tag{3.40}$$

where

$$J_1 = J_b + J_p \frac{r_4}{r_3}\frac{l_4}{l_3} - J_e \frac{r_4(r_6 + \mu r_8)}{r_3(r_5 - \mu r_7)} \frac{l_2 l_4}{l_1 l_3}$$

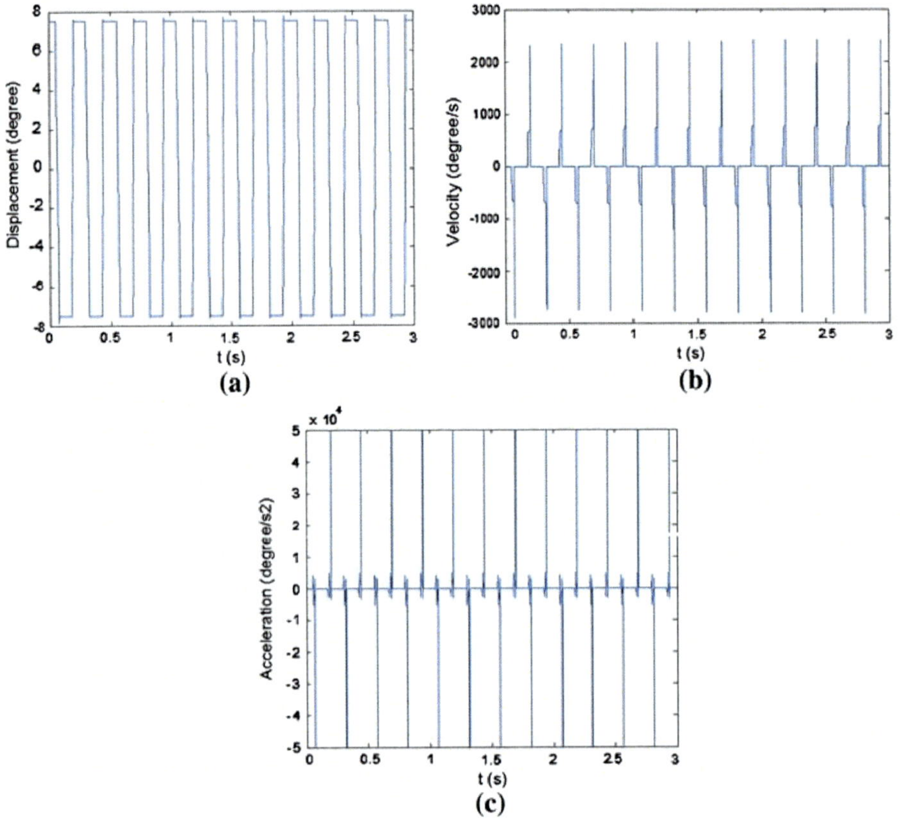

Fig. 3.30 The movement of the pallet fork. **a** The displacement of pallet fork. **b** The velocity of pallet fork. **c** The acceleration of pallet fork

$$T_1 = \tau \frac{r_4(r_6 + \mu r_8)}{r_3(r_5 - \mu r_7)}$$

Here, μ is the friction coefficient.

At the meantime, under the push of the balance wheel, the pallet fork moves following Eq. (3.34). Furthermore, the pallet fork pushes the escape wheel, which recoils a small angle. The escape wheel follows Eq. (3.35).

While there is no shock, this state ends when the pallet fork disengaged from the escape wheel at t_{1p}, giving the displacements and velocities $\theta_{b1p}, \theta_{p1p}, \theta_{e1p}, \dot{\theta}_{b1p}, \dot{\theta}_{p1p}, \dot{\theta}_{e1p}$.

State 4 (P_2, $t = (t_{1p} \sim t_2)$): The escape wheel releases from the pallet fork and continues to rotate under only external torque τ.

$$J_e \ddot{\theta}_e = -\tau \tag{3.41}$$

On the other hand, the balance wheel continues to turn counterclockwise and hence, back to the similar model as Eq. (3.32):

$$J_2\ddot{\theta}_b + K\theta_b = 0 \tag{3.42}$$

Here,

$$J_2 = J_b + J_p\frac{r_4}{r_3}\frac{l_4}{l_3} \tag{3.43}$$

At the same time, the pallet fork continues to move following Eq. (3.34).

This step ends when the escape wheel tooth reaches the pallet again at the moment t_2 giving the displacements and velocities $\theta_{b2}, \theta_{p2}, \theta_{e2}, \dot{\theta}_{b20}, \dot{\theta}_{p20}, \dot{\theta}_{e20}$.

(c) Impulse phase

State 5 (S_2, $t = t_2$): The escape wheel hits the pallet giving the second shock and almost simultaneously, the other side of the pallet fork notch hits the impulse pin as the third shock. In our model, these two shocks are considered to happen at the same time.

$$\frac{J_b}{r_2r_4}\dot{\theta}_{b20} - \frac{J_p}{r_2r_3}\dot{\theta}_{p20} - \sigma\frac{J_e}{r_1r_4}\dot{\theta}_{e20} = \frac{J_b}{r_2r_4}\dot{\theta}_{b21} - \frac{J_p}{r_2r_3}\dot{\theta}_{p21} - \sigma\frac{J_e}{r_1r_4}\dot{\theta}_{e21} \tag{3.44}$$

Because

$$\frac{\dot{\theta}_{b21}}{\dot{\theta}_{p21}} = -\frac{l_3}{l_4} \tag{3.45}$$

$$\frac{\dot{\theta}_{p21}}{\dot{\theta}_{e21}} = \sigma\frac{l_1}{l_2} \tag{3.46}$$

Equation (3.44) can be written as follows:

$$\begin{aligned}\frac{J_b}{r_2r_4}\dot{\theta}_{b20} - \frac{J_p}{r_2r_3}\dot{\theta}_{p20} - \sigma\frac{J_e}{r_1r_4}\dot{\theta}_{e20} &= \left(\frac{J_b}{r_2r_4} + \frac{J_pl_4}{r_2r_3l_3} + \frac{J_el_2l_4}{r_1r_4l_1l_3}\right)\dot{\theta}_{b21}\\ &= -\left(\frac{J_bl_3}{r_2r_4l_4} + \frac{J_p}{r_2r_3} + \frac{J_el_2}{r_1r_4l_1}\right)\dot{\theta}_{p21}\\ &= \sigma\left(\frac{J_bl_1l_3}{r_2r_4l_2l_4} + \frac{J_pl_1}{r_2r_3l_2} + \frac{J_e}{r_1r_4}\right)\dot{\theta}_{e21}\end{aligned} \tag{3.47}$$

Also,

$$\begin{aligned}\Delta\dot{\theta}_b =& \frac{J_er_2r_3l_2l_4}{J_br_1r_3l_1l_3 + J_pr_1r_4l_1l_4 + J_er_2r_3l_2l_4}\dot{\theta}_{b20}\\ &- \frac{\sigma J_er_2r_3l_1l_3}{J_br_1r_3l_1l_3 + J_pr_1r_4l_1l_4 + J_er_2r_3l_2l_4}\dot{\theta}_{e20}\end{aligned} \tag{3.48}$$

$$\begin{aligned}\Delta\dot{\theta}_p =& -\frac{J_er_2r_3l_2l_4}{J_br_1r_3l_1l_3 + J_pr_1r_4l_1l_4 + J_er_2r_3l_2l_4}\frac{l_4}{l_3}\dot{\theta}_{b20}\\ &+ \frac{\sigma J_er_2r_3l_1l_4}{J_br_1r_3l_1l_3 + J_pr_1r_4l_1l_4 + J_er_2r_3l_2l_4}\dot{\theta}_{e20}\end{aligned} \tag{3.49}$$

$$
\begin{aligned}
\Delta\dot{\theta}_e = {} & \sigma \frac{J_e r_2 r_3 l_2 l_4}{J_b r_1 r_3 l_1 l_3 + J_p r_1 r_4 l_1 l_4 + J_e r_2 r_3 l_2 l_4} \frac{l_2 l_4}{l_1 l_3} \dot{\theta}_{b20} \\
& - \sigma \frac{\sigma J_e r_2 r_3 l_2 l_4}{J_b r_1 r_3 l_1 l_3 + J_p r_1 r_4 l_1 l_4 + J_e r_2 r_3 l_2 l_4} \dot{\theta}_{e20}
\end{aligned}
\tag{3.50}
$$

State 6 (P_3, $t = (t_2 \sim t_3)$): The escape wheel pushes the pallet fork and balance wheel together and restores the energy exhausted during the previous impulses. For the balance wheel,

$$
J_3 \ddot{\theta}_b + K\theta_b = \sigma T_3 \tag{3.51}
$$

where

$$
J_3 = J_b + J_p \frac{r_4}{r_3} \frac{l_4}{l_3} + J_e \frac{(r_2 - \mu r_{10}) r_4}{(r_1 + \mu r_9) r_3} \frac{l_2 l_4}{l_1 l_3}
$$

$$
T_3 = \tau \frac{(r_2 - \mu r_{10}) r_4}{(r_1 + \mu r_9) r_3}
$$

The pallet fork and the escape wheel follow their previous equations (Eqs. 3.45 and 3.46), respectively.

This step ends when the escape tooth leaves the pallet fork giving the displacements and velocities $\theta_{b3}, \theta_{p3}, \theta_{e3}, \dot{\theta}_{b3}, \dot{\theta}_{p3}, \dot{\theta}_{e3}$.

State 7 (P_4, $t = (t_3 \sim t_4)$): This step is almost the same as State 4. For the balance wheel,

$$
J_2 \ddot{\theta}_b + K\theta_b = 0 \tag{3.52}
$$

For the escape wheel,

$$
J_e \ddot{\theta}_e = -\tau \tag{3.53}
$$

The motion of the pallet fork is unchanged.

This step ends when one escape wheel tooth reaches the other pallet giving the displacements and velocities $\theta_{b4}, \theta_{p4}, \theta_{e4}, \dot{\theta}_{b40}, \dot{\theta}_{p40}, \dot{\theta}_{e40}$

(d) Safety phase

State 8 (S_3, $t = t_4$): At this time, the escape wheel hits the pallet fork giving the fourth shock. This state is similar to State 5:

$$
\frac{J_b}{r_4 r_6} \dot{\theta}_{b40} - \frac{J_p}{r_3 r_6} \dot{\theta}_{p40} - \sigma \frac{J_e}{r_4 r_5} \dot{\theta}_{e40} = \frac{J_b}{r_4 r_6} \dot{\theta}_{b41} - \frac{J_p}{r_3 r_6} \dot{\theta}_{p41} - \sigma \frac{J_e}{r_4 r_5} \dot{\theta}_{e41} \tag{3.54}
$$

$$
\begin{aligned}
\Delta\dot{\theta}_b = {} & \frac{J_e r_3 r_6 l_2 l_4}{J_b r_3 r_5 l_1 l_3 + J_p r_4 r_5 l_1 l_4 + J_e r_3 r_6 l_2 l_4} \dot{\theta}_{b40} \\
& - \frac{\sigma J_e r_3 r_6 l_1 l_3}{J_b r_3 r_5 l_1 l_3 + J_p r_4 r_5 l_1 l_4 + J_e r_3 r_6 l_2 l_4} \dot{\theta}_{e40}
\end{aligned}
\tag{3.55}
$$

$$\Delta\dot{\theta}_p = -\frac{J_e r_3 r_6 l_2 l_4}{J_b r_3 r_5 l_1 l_3 + J_p r_4 r_5 l_1 l_4 + J_e r_3 r_6 l_2 l_4} \frac{l_4}{l_3} \dot{\theta}_{b40} + \frac{\sigma J_e r_3 r_6 l_1 l_4}{J_b r_3 r_5 l_1 l_3 + J_p r_4 r_5 l_1 l_4 + J_e r_3 r_6 l_2 l_4} \dot{\theta}_{e40} \quad (3.56)$$

$$\Delta\dot{\theta}_e = \sigma \frac{J_e r_3 r_6 l_2 l_4}{J_b r_3 r_5 l_1 l_3 + J_p r_4 r_5 l_1 l_4 + J_e r_3 r_6 l_2 l_4} \frac{l_2 l_4}{l_1 l_3} \dot{\theta}_{b40} - \sigma \frac{\sigma J_e r_3 r_6 l_2 l_4}{J_b r_3 r_5 l_1 l_3 + J_p r_4 r_5 l_1 l_4 + J_e r_3 r_6 l_2 l_4} \dot{\theta}_{e40} \quad (3.57)$$

State 9 (P_5, $t = (t_4 \sim t_5)$): The torque on the escape wheel drives it to impact the pallet fork:

$$J_4 \ddot{\theta}_e = -\tau \quad (3.58)$$

where

$$J_4 = J_e + J_p \frac{(r_5 + \mu r_7)}{(r_6 - \mu r_8)} \frac{l_1}{l_2}$$

The pallet fork continues to follow the previous equation. The balance wheel starts the descending supplementary arc

$$J_b \ddot{\theta}_b + K\theta_b = 0 \quad (3.59)$$

This step ends when the pallet fork reaches the banking pin giving the displacements and velocities $\theta_{b5}, \theta_{p5}, \theta_{e5}, \dot{\theta}_{b50}, \dot{\theta}_{p50}, \dot{\theta}_{e50}$. Note that

$$\theta_{b5} = -\theta_{b1} \quad (3.60)$$

State 10 (S_4, $t = t_5$): The fifth, the last, shock occurs when the pallet fork collides onto the banking pin

$$\dot{\theta}_{p51} = 0, \dot{\theta}_{e51} = 0 \quad (3.61)$$

Therefore,

$$\Delta\dot{\theta}_b = 0 \quad (3.62)$$

$$\Delta\dot{\theta}_p = -\dot{\theta}_{e50} \quad (3.63)$$

$$\Delta\dot{\theta}_e = -\dot{\theta}_{e50} \quad (3.64)$$

At this time, both the escape wheel and the pallet fork are locked; though the balance wheel continues to swing following the previous equation.

(a′) Lock phase

State 11 (P_6, $t = (t_5 \sim t_6)$): This is the lock phase of the second half of the cycle. The balance wheel continues the descending along the supplementary arc while the pallet fork and the escape wheel are locked:

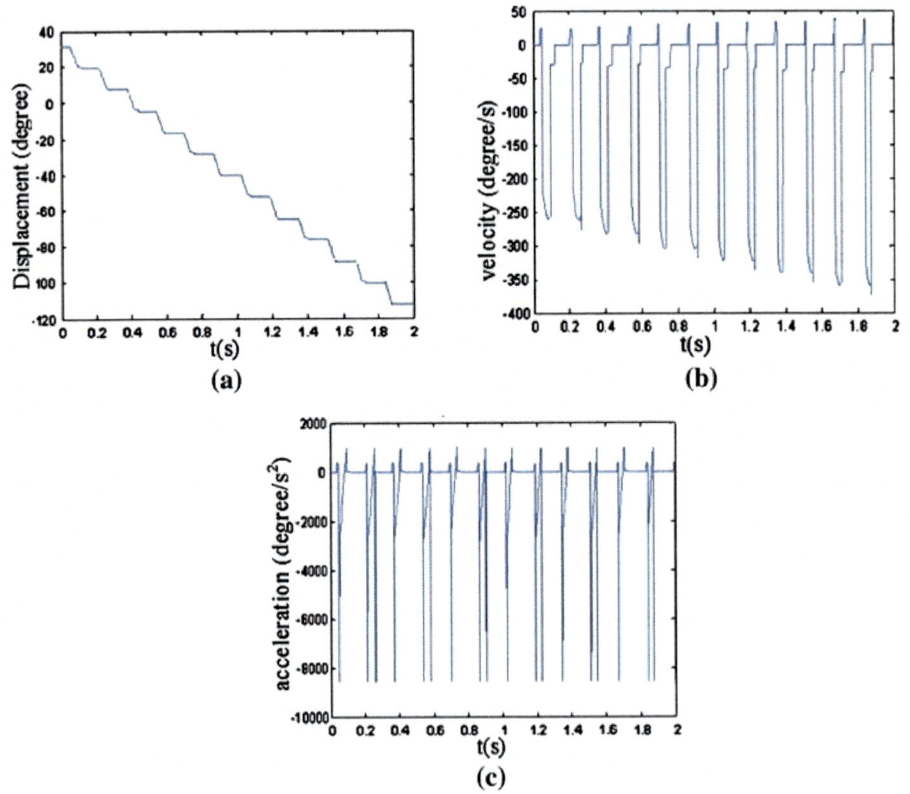

Fig. 3.31 The movement of the escape wheel. **a** The displacement of escape wheel. **b** The velocity of escape wheel. **c** The acceleration

$$J_b\ddot{\theta}_b + K\theta_b = 0 \tag{3.65}$$

This step ends when the angle reaches the maximum θ_{b5}, on which the angular velocity of the balance wheel is zero. Afterwards, the balance wheel goes back in the opposite direction and the movement is symmetric.

In summary, Fig. 3.25 gives the timeline of the 11 states and the changing of the equations. In the figure, the *arrow* indicates the states and the *dash line* represents the transmission of the torque. From the figure, it is seen that the Swiss lever escapement changes its movement 11 times in half a cycle, making it one of the most sophisticated mechanical systems ever invented.

Now, we can assemble the model of the Swiss lever escapement in the half cycle using the impulsive differential equation. Define the state:

$$x \equiv [\theta_b, \theta_p, \theta_e, \dot{\theta}_b, \dot{\theta}_p, \dot{\theta}_e]^T \tag{3.66}$$

The jump set is

$$S = \{x : \varphi(x) = 0\} = S_1 \cup S_2 \cup S_3 \cup S_4 \tag{3.67}$$

in which $S_1 = \{x : x_1 = \theta_{b1}\}$, $S_2 = \{x : x_1 = \theta_{b2}\}$, $S_3 = \{x : x_1 = \theta_{b4}\}$ and $S_4 = \{x : x_1 = \theta_{b5}\}$ are jump points. Accordingly, the impulsive differential equation and the jump function are given by

$$\dot{x} = f(x)$$
$$= \begin{cases} \begin{bmatrix} 0 & 0 & 0 & 1 & 0 & 0 \\ 0 & 0 & 0 & 0 & 1 & 0 \\ 0 & 0 & 0 & 0 & 0 & 1 \\ -\frac{K}{J_b} & 0 & 0 & 0 & 0 & 0 \\ 0 & 0 & 0 & 0 & 0 & 0 \\ 0 & 0 & 0 & 0 & 0 & 0 \end{bmatrix} x, & x \in P_1 \cup P_6 \\ \begin{bmatrix} 0 & 0 & 0 & 1 & 0 & 0 \\ 0 & 0 & 0 & 0 & 1 & 0 \\ 0 & 0 & 0 & 0 & 0 & 1 \\ -\frac{K}{J_1} & 0 & 0 & 0 & 0 & 0 \\ 0 & 0 & 0 & 0 & 0 & 0 \\ 0 & 0 & 0 & 0 & 0 & 0 \end{bmatrix} x + \begin{bmatrix} 0 \\ 0 \\ 0 \\ -o\frac{1}{J_1}\frac{r_4(r_6+\mu r_8)}{r_3(r_5-\mu r_7)} \\ o\frac{1}{J_1}\frac{r_4(r_6+\mu r_8)}{r_3(r_5-\mu r_7)}\frac{l_4}{l_3} \\ \frac{1}{J_1}\frac{r_4(r_6+\mu r_8)}{r_3(r_5-\mu r_7)}\frac{l_2 l_4}{l_1 l_3} \end{bmatrix} \tau, & x \in P_{1p} \\ \begin{bmatrix} 0 & 0 & 0 & 1 & 0 & 0 \\ 0 & 0 & 0 & 0 & 1 & 0 \\ 0 & 0 & 0 & 0 & 0 & 1 \\ -\frac{K}{J_2} & 0 & 0 & 0 & 0 & 0 \\ 0 & -\frac{K}{J_2} & 0 & 0 & 0 & 0 \\ 0 & 0 & 0 & 0 & 0 & 0 \end{bmatrix} x + \begin{bmatrix} 0 \\ 0 \\ 0 \\ 0 \\ 0 \\ -\frac{1}{J_e} \end{bmatrix} \tau, & x \in P_2 \cup P_4 \\ \begin{bmatrix} 0 & 0 & 0 & 1 & 0 & 0 \\ 0 & 0 & 0 & 0 & 1 & 0 \\ 0 & 0 & 0 & 0 & 0 & 1 \\ -\frac{K}{J_3} & 0 & 0 & 0 & 0 & 0 \\ 0 & -\frac{K}{J_3} & 0 & 0 & 0 & 0 \\ o\frac{K}{J_3}\frac{l_2 l_4}{l_1 l_3} & 0 & 0 & 0 & 0 & 0 \end{bmatrix} x + \begin{bmatrix} 0 \\ 0 \\ 0 \\ o\frac{1}{J_3}\frac{(r_2-\mu \mathrm{r}_{10})r_4}{(r_1+\mu \mathrm{r}_9)r_3} \\ -o\frac{1}{J_3}\frac{(r_2-\mu \mathrm{r}_{10})r_4}{(r_1+\mu \mathrm{r}_9)r_3}\frac{l_4}{l_3} \\ -\frac{1}{J_3}\frac{(r_2-\mu \mathrm{r}_{10})r_4}{(r_1+\mu \mathrm{r}_9)r_3}\frac{l_2 l_4}{l_1 l_3} \end{bmatrix} \tau, & x \in P_3 \\ \begin{bmatrix} 0 & 0 & 0 & 1 & 0 & 0 \\ 0 & 0 & 0 & 0 & 1 & 0 \\ 0 & 0 & 0 & 0 & 0 & 1 \\ -\frac{K}{J_b} & 0 & 0 & 0 & 0 & 0 \\ 0 & 0 & 0 & 0 & 0 & 0 \\ 0 & 0 & 0 & 0 & 0 & 0 \end{bmatrix} x + \begin{bmatrix} 0 \\ 0 \\ 0 \\ 0 \\ -o\frac{1}{J_4}\frac{l_1}{l_2} \\ -\frac{1}{J_4} \end{bmatrix} \tau, & x \in P_5 \end{cases} \tag{3.68}$$

$$\rho(x) = \begin{cases} \begin{bmatrix} 0 & 0 & 0 & 0 & 0 & 0 \\ 0 & 0 & 0 & 0 & 0 & 0 \\ 0 & 0 & 0 & 0 & 0 & 0 \\ 0 & 0 & 0 & a_1 & 0 & 0 \\ 0 & 0 & 0 & a_2 & 0 & 0 \\ 0 & 0 & 0 & a_3 & 0 & 0 \end{bmatrix} x, & x \in S_1 \\ \begin{bmatrix} 0 & 0 & 0 & 0 & 0 & 0 \\ 0 & 0 & 0 & 0 & 0 & 0 \\ 0 & 0 & 0 & 0 & 0 & 0 \\ 0 & 0 & 0 & a_4 & 0 & a_5 \\ 0 & 0 & 0 & a_6 & 0 & a_7 \\ 0 & 0 & 0 & a_8 & 0 & a_9 \end{bmatrix} x, & x \in S_2 \\ \begin{bmatrix} 0 & 0 & 0 & 0 & 0 & 0 \\ 0 & 0 & 0 & 0 & 0 & 0 \\ 0 & 0 & 0 & 0 & 0 & 0 \\ 0 & 0 & 0 & a_{10} & 0 & a_{11} \\ 0 & 0 & 0 & a_{12} & 0 & a_{13} \\ 0 & 0 & 0 & a_{14} & 0 & a_{15} \end{bmatrix} x, & x \in S_3 \\ \begin{bmatrix} 0 & 0 & 0 & 0 & 0 & 0 \\ 0 & 0 & 0 & 0 & 0 & 0 \\ 0 & 0 & 0 & 0 & 0 & 0 \\ 0 & 0 & 0 & 0 & 0 & 0 \\ 0 & 0 & 0 & 0 & -1 & 0 \\ 0 & 0 & 0 & 0 & 0 & -1 \end{bmatrix} x, & x \in S_4 \end{cases} \tag{3.69}$$

where

$$a_1 = -\frac{J_p r_4 r_5 l_1 l_4 + J_e r_4 r_6 l_2 l_4}{J_b r_3 r_5 l_1 l_3 + J_p r_4 r_5 l_1 l_4 + J_e r_4 r_6 l_2 l_4}$$

$$a_2 = -\frac{J_b r_3 r_5 l_1 l_4}{J_b r_3 r_5 l_1 l_3 + J_p r_4 r_5 l_1 l_4 + J_e r_4 r_6 l_2 l_4}$$

$$a_3 = \sigma \frac{J_b r_3 r_5 l_2 l_4}{J_b r_3 r_5 l_1 l_3 + J_p r_4 r_5 l_1 l_4 + J_e r_4 r_6 l_2 l_4}$$

$$a_4 = \frac{J_e r_2 r_3 l_2 l_4}{J_b r_1 r_3 l_1 l_3 + J_p r_1 r_4 l_1 l_4 + J_e r_2 r_3 l_2 l_4}$$

$$a_5 = -\frac{\sigma J_e r_2 r_3 l_1 l_3}{J_b r_1 r_3 l_1 l_3 + J_p r_1 r_4 l_1 l_4 + J_e r_2 r_3 l_2 l_4}$$

$$a_6 = -\frac{J_e r_2 r_3 l_2 l_4}{J_b r_1 r_3 l_1 l_3 + J_p r_1 r_4 l_1 l_4 + J_e r_2 r_3 l_2 l_4}$$

$$a_7 = \frac{\sigma J_e r_2 r_3 l_1 l_4}{J_b r_1 r_3 l_1 l_3 + J_p r_1 r_4 l_1 l_4 + J_e r_2 r_3 l_2 l_4}$$

$$a_8 = \sigma \frac{J_e r_2 r_3 l_2 l_4}{J_b r_1 r_3 l_1 l_3 + J_p r_1 r_4 l_1 l_4 + J_e r_2 r_3 l_2 l_4} \frac{l_2 l_4}{l_1 l_3}$$

$$a_9 = -\sigma \frac{\sigma J_e r_2 r_3 l_2 l_4}{J_b r_1 r_3 l_1 l_3 + J_p r_1 r_4 l_1 l_4 + J_e r_2 r_3 l_2 l_4}$$

$$a_{10} = \frac{J_e r_3 r_6 l_2 l_4}{J_b r_3 r_5 l_1 l_3 + J_p r_4 r_5 l_1 l_4 + J_e r_3 r_6 l_2 l_4}$$

$$a_{11} = -\frac{\sigma J_e r_3 r_6 l_1 l_3}{J_b r_3 r_5 l_1 l_3 + J_p r_4 r_5 l_1 l_4 + J_e r_3 r_6 l_2 l_4}$$

$$a_{12} = -\frac{J_e r_3 r_6 l_2 l_4}{J_b r_3 r_5 l_1 l_3 + J_p r_4 r_5 l_1 l_4 + J_e r_3 r_6 l_2 l_4} \frac{l_4}{l_3}$$

$$a_{13} = \frac{\sigma J_e r_3 r_6 l_1 l_4}{J_b r_3 r_5 l_1 l_3 + J_p r_4 r_5 l_1 l_4 + J_e r_3 r_6 l_2 l_4}$$

$$a_{14} = \sigma \frac{J_e r_3 r_6 l_2 l_4}{J_b r_3 r_5 l_1 l_3 + J_p r_4 r_5 l_1 l_4 + J_e r_3 r_6 l_2 l_4} \frac{l_2 l_4}{l_1 l_3}$$

$$a_{15} = -\sigma \frac{\sigma J_e r_3 r_6 l_2 l_4}{J_b r_3 r_5 l_1 l_3 + J_p r_4 r_5 l_1 l_4 + J_e r_3 r_6 l_2 l_4}$$

The supplementary set is

$$P = \bar{S} = P_1 \cup P_{1p} \cup P_2 \cup P_3 \cup P_4 \cup P_5 \cup P_6 \tag{3.70}$$

where

$$P_1 = \{x : x_1 > \theta_{b1}\}$$

$$P_{1p} = \{x : (x_1 - \theta_{b1})(x_1 - \theta_{b1p}) < 0\}$$

$$P_2 = \{x : (x_1 - \theta_{b1p})(x_1 - \theta_{b2}) < 0\}$$

$$P_3 = \{x : (x_1 - \theta_{b2})(x_1 - \theta_{b3}) < 0\}$$

$$P_4 = \{x : (x_1 - \theta_{b3})(x_1 - \theta_{b4}) < 0\}$$

$$P_5 = \{x : (x_1 - \theta_{b4})(x_1 - \theta_{b5}) < 0\}$$

$$P_6 = \{x : x_1 < \theta_{b5}\}$$

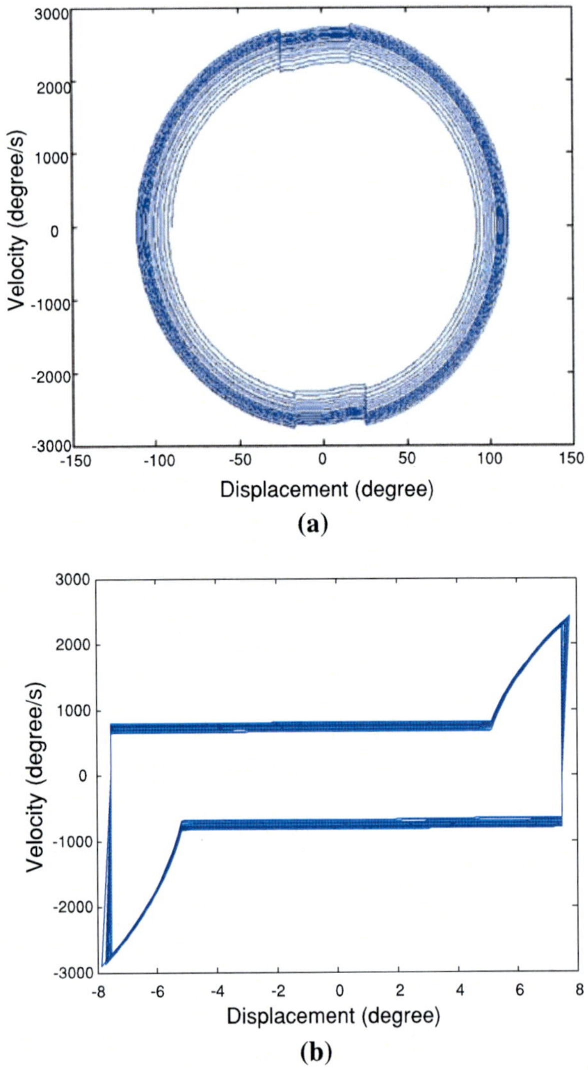

Fig. 3.32 The phase diagrams of the balance wheel and the pallet fork. **a** The balance wheel. **b** The pallet fork

The equation above can be numerically solved giving the displacement, velocity and acceleration of the escape wheel, the pallet fork and the balance wheel of the Swiss lever mechanism.

In particularly, two sets of simulations are presented below.

(a) *Simulation 1:* Figures 3.26, 3.27 and 3.28 show the time history of the angular displacement, velocity, acceleration of the balance wheel, the pallet fork and the escape wheel in one cycle. From the figures, several observations can be made. First, it is interesting to compare Figs. 3.21 and 3.26: The former is calculated based on the geometry, whereas the latter is based on dynamics. It is seen that both result in the same pattern.

Fig. 3.33 The torque on the escape wheel versus the magnitude of oscillation

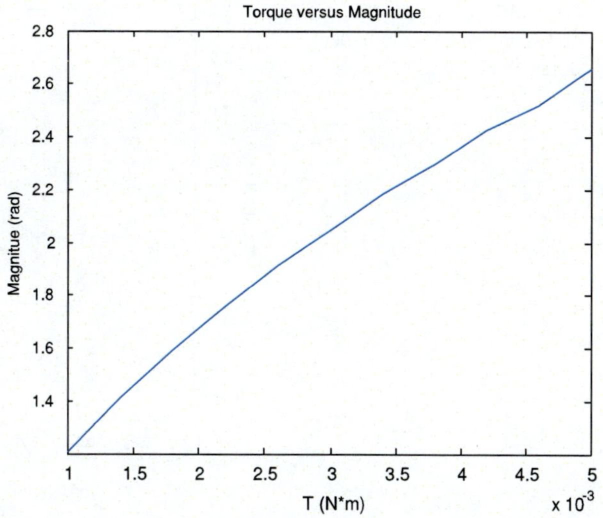

Fig. 3.34 θ_{b0} versus the oscillation magnitude under different torques

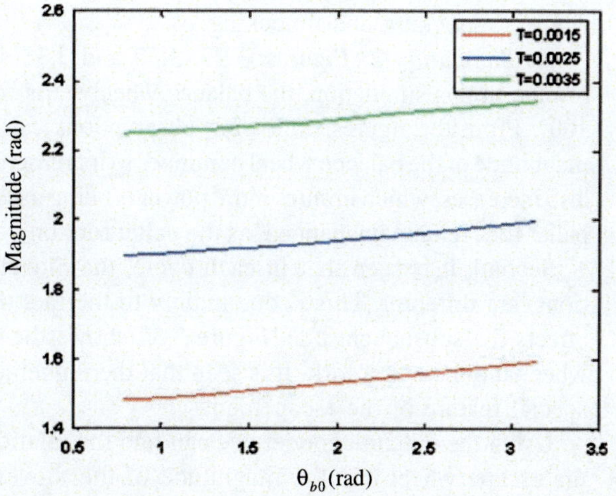

Second, from Fig. 3.27, it is seen that the balance wheel, the pallet fork and the escape wheel all experience rapid velocity changes. In particularly, the first and the fifth shocks are critical for the pallet fork, whereas the second and the fourth shocks are most important for escape wheel. Examining the timeline in Fig. 3.25, the first and the fifth shocks correspond to the pallet fork impacting the balance wheel and the blanking pin. The second and the fourth shock correspond to the escape wheel impacting the pallet fork.

Third, from Fig. 3.28, it is seen that the balance wheel receives the two push in a cycle, at which time the balance wheel accelerates. The acceleration of the pallet

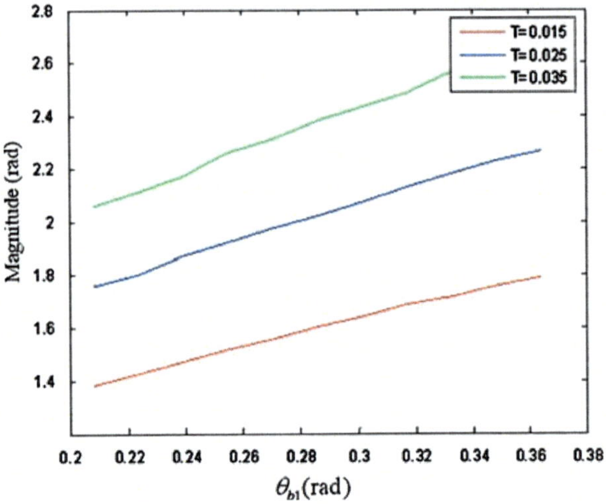

Fig. 3.35 θ_{b1} versus the oscillation magnitude under different torques

fork is complex, as it transfers the motion and the energy between the balance wheel and the escape wheel. The acceleration of the escape wheel is also complex due to the locking and unlocking.

(b) *Simulation 2:* Figures. 3.29, 3.30 and 3.31 show another set of simulation results. In this simulation, the balance wheel starts from a small oscillation angle of 100°. From the figures, following observations can be made: First, the oscillation magnitude of the balance wheel continues to increase. The velocity of the escape wheel also increases, which implies more power is being used. Though, the movements of the pallet fork remain unchanged, as the pallet fork only transfers the power.

Second, it is seen that in each cycle, the movements, especially the accelerations, are different. This is due mainly to the fact that the dynamics in each cycle affects the subsequent one. Figure 3.32 shows the phase diagrams of the balance wheel and the pallet fork. It is seen that the movement is almost periodic. This is a special feature of the escapement.

Using the dynamic model, we can find the relationship between the torque from the escape wheel and the magnitude of the movement of the balance wheel, as shown in Fig. 3.33. The proportional relation is clearly seen. It is known that large magnitude implies more stored energy and therefore can minimize the effect of the disturbances, such as gravity and temperature. Therefore, larger torque from the escape wheel is desirable.

Figures 3.34 and 3.35 show the influence of θ_{b0} (the initial angle of the balance wheel) and θ_{b1} (the unlocking angle of the balance wheel) to the oscillation magnitude of the balance wheel under different torques. They both exhibit positive correlation. But in Fig. 3.34, the slope is small, which implies that the effect of the initial angle of the balance wheel is small. On the other hand, in Fig. 3.35, the slope is large, which implies the unlocking angle of the balance angle is sensitive.

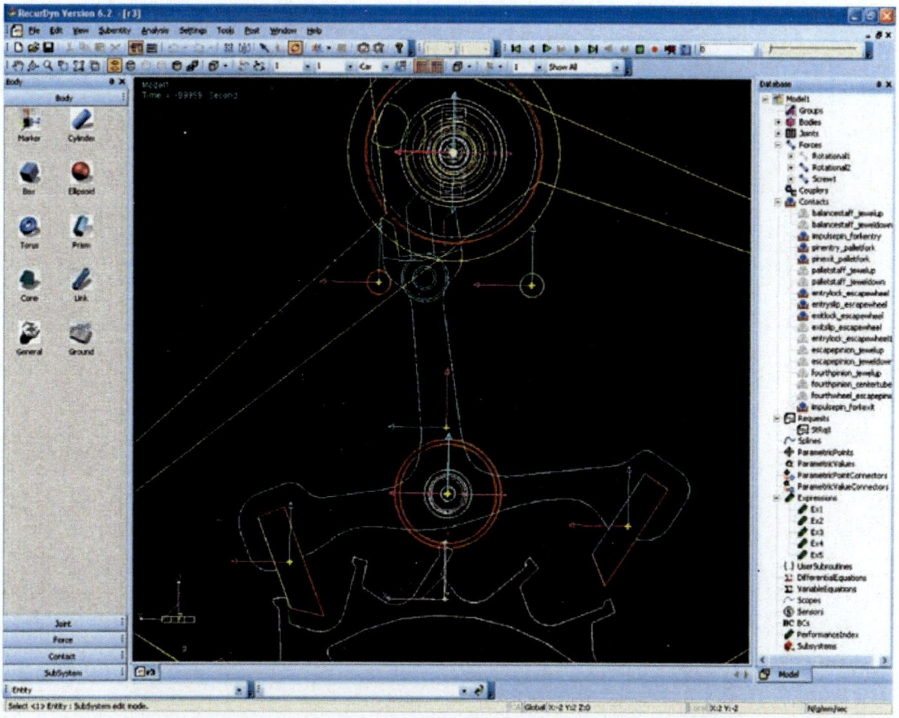

Fig. 3.36 The Swiss lever escapement model in RecurDyn

3.4 Modeling Using RecurDyn®

In recent years, computer software systems have been greatly advanced. There are special software systems that can compute the dynamics of the complex systems, such as Adams Multibody Dynamics by MSC Software and RecurDyn Multi-Body Simulation and FEA by Functionbay. We use RecurDyn to model and simulate the Swiss lever escapement. Figure 3.36 shows the model in RecurDyn. The difficulty comes when setting the boundary conditions in the shocks. After many trials, a compromise is used: It is assumed that the balance wheel swings following a pre-set sinusoidal trajectory (as that of in Sect. 3.2). Accordingly, the movements are obtained. Figure 3.37 shows the angular displacement of the balance wheel, the pallet fork and the escape wheel. In comparison to Fig. 3.23, the simulation results from RecurDyn have a strange peak in the angular displacements of the pallet fork and the escape wheel. In addition, in the angular displacement of the escape wheel, the escape wheel recoil is not seen. These are believed to be errors.

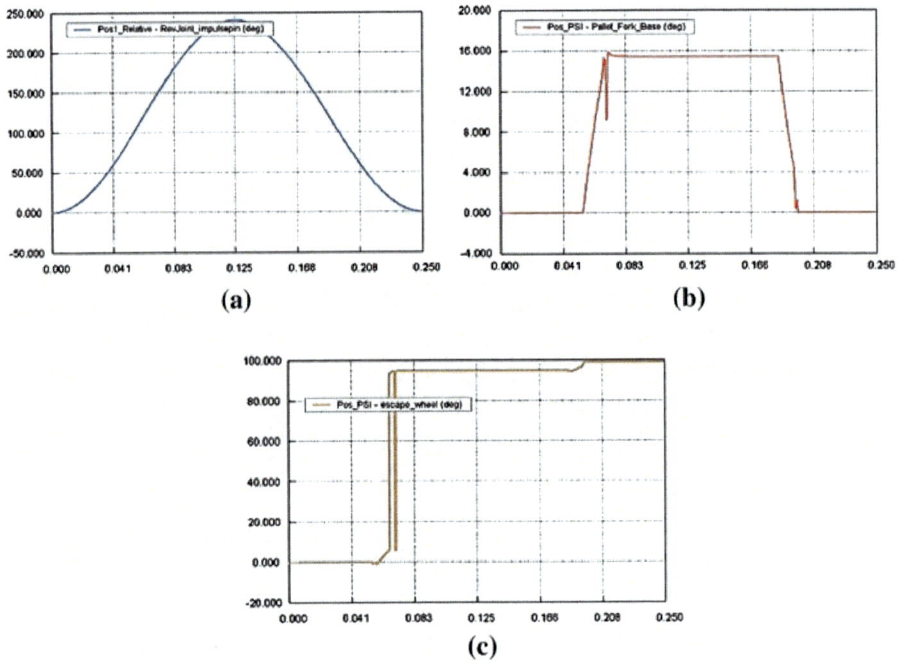

Fig. 3.37 The angular displacement of the Swiss lever escapement. **a** The balance wheel. **b** The pallet fork. **c** The escape wheel

3.5 Experimental Validation Using Acoustic Signals

The presented model is validated by means of experiments. Because of the size and complexity of the escapement, it is very difficult to measure its displacement, or velocity or acceleration. Therefore, we use an indirect method: analyze the sound of the escapement (Su and Du 2007). As pointed out in the beginning of the chapter, the sound is from the vibration. There are commercial systems for testing mechanical watch movements using the sound; however, we cannot extract signals from these systems. Therefore, we developed an audio signal acquisition and analysis system as shown in Fig. 3.38.

Figure 3.39 shows a typical signal. Zooming into one beat (Fig. 3.39c), the aforementioned three-peak pattern can be clearly seen.

In general, the performance of the mechanical watch movement is measured by two parameters: the day deviation, Ω; and the amplitude of the balance wheel, Φ. The former is defined as follow:

$$\Omega = 86400 \frac{\Delta T}{T} \tag{3.71}$$

Fig. 3.38 Our audio signal acquisition system for mechanical watch movements

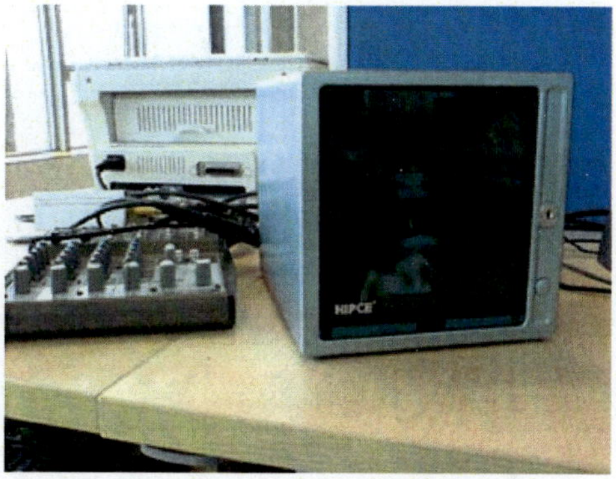

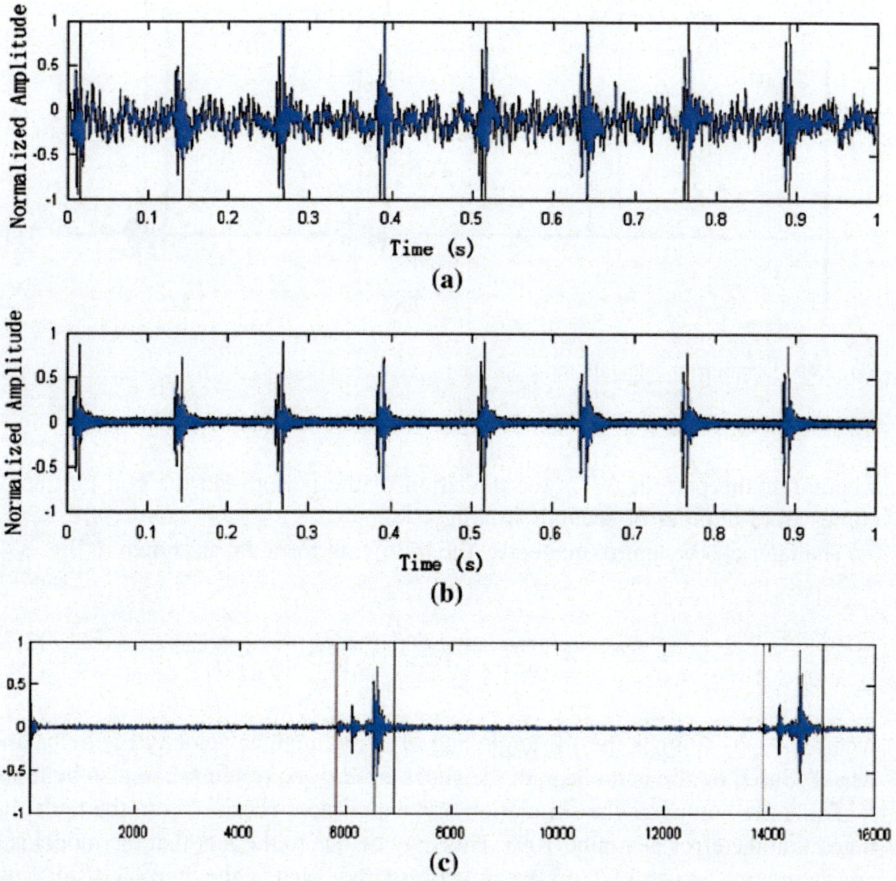

Fig. 3.39 A typical acoustic signal from a mechanical watch movement. **a** The original signal. **b** The signal after low-pass filtering. **c** The signal in a beat

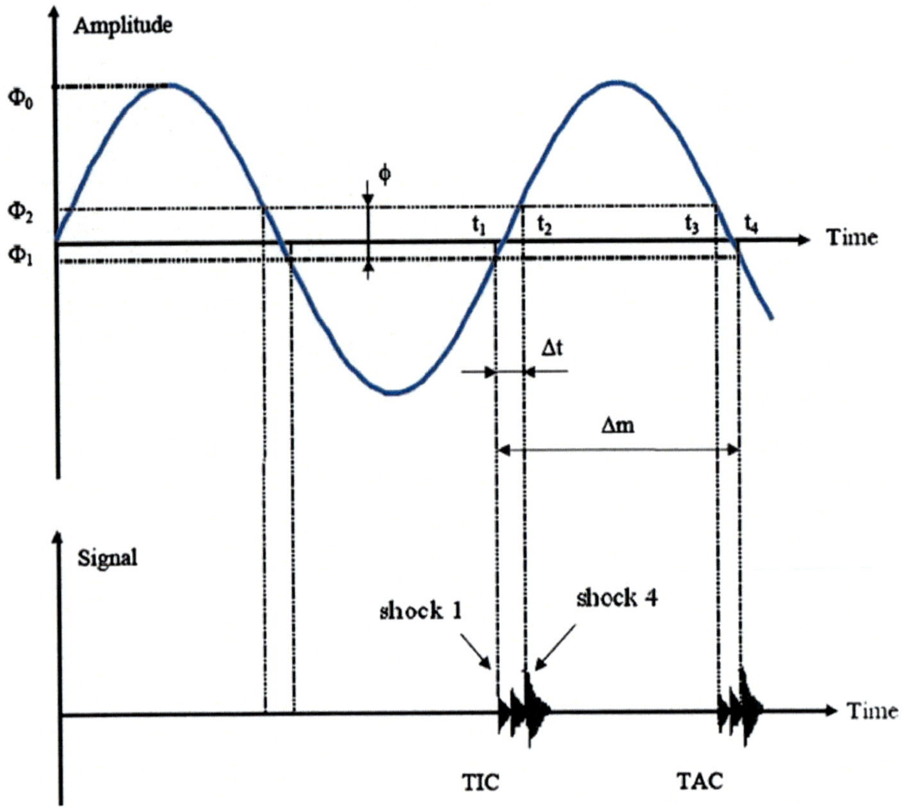

Fig. 3.40 Diagram to calculate the amplitude

where T is the period, ΔT is the deviation of the period and the constant 86400 comes from number of seconds in a day.

The later can be approximated by the following formula, as shown in Fig. 3.40.

$$\Phi_0 = \frac{\beta}{2 \cdot \sin\left[\omega \cdot \left(\frac{\Delta m - \Delta t}{2}\right)\right] \cdot \sin\left[\omega \cdot \left(\frac{\Delta t}{2}\right)\right]} \tag{3.72}$$

where $\beta = \Phi_2 - \Phi_1$ is the lift angle and ω is the angular velocity of the balance wheel. Based on the audio signal, the shock time t_1, t_2, t_3 and t_4 can also be found.

Table 3.1 compares the simulation and experiment results. From the table, it is noted that the error is around 10%. This may be due to the fact that the model is an approximation, several factors are not considered, such as the thermal disturbance, the gravity disturbance, the dimension inaccuracy of the parts, the friction among the parts, the effect of the hairspring, etc. On the other hand, each manufactured mechanical watch movement is slightly different, because of the material inhomogeneity, design tolerance, manufacturing and assembly error, lubrication, etc. As a result, a 10% error is expected.

It may be argued that the above experiments did not really validate the model as it is an indirect measure and the error is significant. We also tried to use laser stroboscopic; unfortunately, the resulting error is about the same (Wang et al. 2008).

3.6 Sensitivity Analysis

Based on the model presented above, we can study how the geometric parameters affect the performance of the escapement. This will help the design and manufacturing quality. Figure 3.41 shows four key manufacturing dimensions D1, D2, D3 and D4.

Tables 3.2 and 3.3 show the simulation results under different manufacturing tolerance. It is seen that D3 has the largest effect on the time keeping accuracy. Therefore, this dimension should be tightly controlled in the manufacturing process to get better quality.

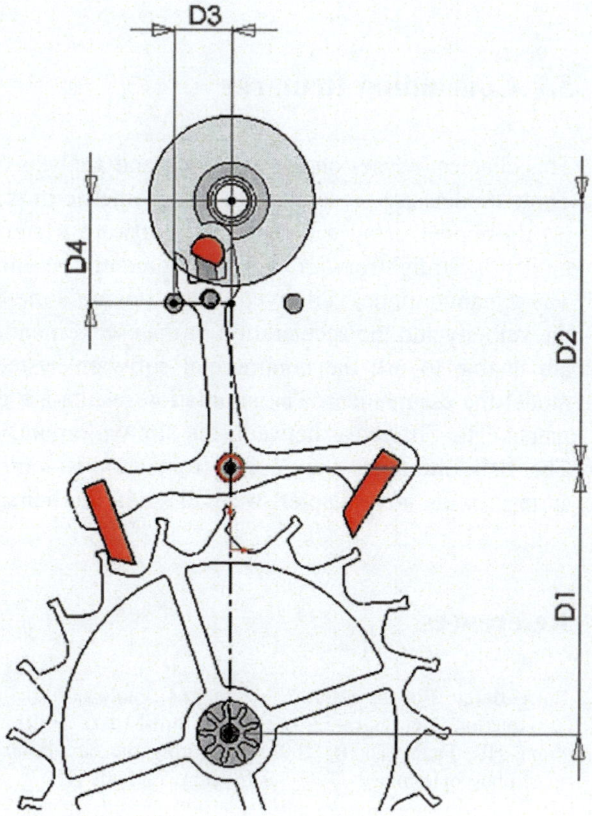

Fig. 3.41 Four basic dimensions in escapement

Table 3.1 Comparison of day deviation (s/day)

	Simulation	Experimental	Error (%)
Day deviation (s/day)	9.3	8.23	11.5
Amplitude (°)	221	242	−9.5

Table 3.2 Day deviation (s/day) in the sensitivity analysis

Tolerance (μm)	±1	±2	±3
D1	1.3	2.5	3.8
D2	0.8	1.6	2.5
D3	3.8	7.5	11.2
D4	1.7	3.3	5.0

Table 3.3 Amplitude deviation (°) in the sensitivity analysis

Tolerance (μm)	±1	±2	±3
D1	0.02	0.04	0.05
D2	0.01	0.02	0.03
D3	0.05	0.10	0.15
D4	0.02	0.04	0.07

3.7 Concluding Remarks

This chapter focuses on the modeling and analysis of the Swiss lever escapement. Three models are presented: (a) the geometric model, (b) the dynamic model and (c) the model using a commercial software system RecurDyn®. The geometric model is straightforward and can predict the movements of the escapement. The dynamic model is derived using Newton's method. It gives the displacement, the velocity and the acceleration of the escapement in good accuracy. Though, we are unable to use the commercial software system Recurdyn® to successfully model the escapement. The simulation results are evaluated by means of experiments. The difference between the simulation and the experiment is about 10%. The 10% difference may come from various sources, one of which is the hairspring. In the next chapter, we will study the hairspring in detail.

References

Borri Marco, Bottassso Carlo, Mantegazza Paolo (1990) Equivalence of Kane's and Maggi's Equation. Meccanica 25(4):272–274. doi:10.1007/FB01559692

Wang HR, Fu Y, Du R (2008) Measuring the circular motion of small objects using laser stroboscopic images, Rev Sci Instrum, 79(1):015110

Kauderer H (1958) Nichtlineare mechanik, (English Translation: Nonlinear Mechanics). Springer, Berlin

Lakshmikantham V, Bainov D.D, Simeonov P.S (1989) Theory of impulsive differential equation. World Scientific Publishing Co. Pte. Ltd, Singapore

Manuel D.P, Marques M, Marques M (1993) Differential inclusions in nonsmooth mechanical problems: shocks and dry friction, progress in nonlinear differential equations and their applications. Birkhauser Publishing, Basel

Roup A.V et al. (2001) Limit cycle analysis of the verge and Foliot clock escapement using impulsive differential equations and poincare maps, In: Proceedings of the 2001 American control conference, (4):3245–3250

Reymondin CA, Monnier G, Jeanneret D, Pelaratti U (1999) The theory of horology. The Technical College of Vallee de Joux, Switzerland

Su S, Du R (2007) Signature analysis of mechanical watch movements. Mech Syst Signal Process 21(8):3189–3200

Gran R, Schwatz C (2011) How to build a clock or controlling an oscillation in nonlinear system using MATLAB simulink, and the control system toolbox, http://www.mathworks.com/company/newsletters/digest/june99/clock/. Accessed 12 Dec 2011

Chapter 4
The Mechanics of the Spiral Spring

When working with the English watch master Thomas Thompin in 1678, Robert Hook (1635–1703) observed that when an elastic body is subjected to stress, its dimension or shape changes in proportion to the applied stress over a range of stresses. On the basis of his experiments with springs, stretching wires and coils, he discovered a relationship between the force and the extension of the spring. This is the so-called Hooke's law which states that strain, the relative change in dimension, is proportional to stress. If the stress applied to a body goes beyond a certain value known as the elastic limit, the body does not return to its original state once the stress is removed. In other words, the Hooke's law applies only in the region below the elastic limit. Mathematically, Hooke's law has the following form:

$$F = -kx \tag{4.1}$$

where, F is the applied force, k is the spring constant and x is displacement. Since then, more than three hundred years has passed, now the Hooke's law is taught in every elementary school in the world.

Nowadays, springs are used everywhere, including, of course, the mechanical watch and clock. This chapter is devoted to the spring: its mechanics and applications in the mechanical watch movements.

4.1 A Historical Review of Spiral Springs, Hairspring and Main Springs

The mechanical watch movement has two springs: the hairspring and the mainspring. They are both spiral springs and extremely important. The hairspring is a fine spiral spring, usually assembled onto the balance wheel to form a harmonic

R. Du and L. Xie, *The Mechanics of Mechanical Watches and Clocks*,
History of Mechanism and Machine Science 21, DOI: 10.1007/978-3-642-29308-5_4,

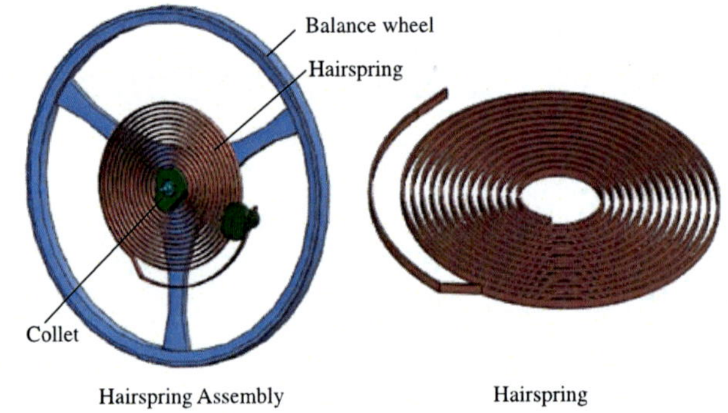

Fig. 4.1 Hairspring and its assembly

oscillator, whose resonant period is a constant. It is a part of the "brain" of the mechanical watch movement (the escapement). The mainspring is a spiral torsion spring of metal ribbon that provides power for the mechanical watch movement. It can be considered as the "heart" of the mechanical watch movement.

As shown in Fig. 4.1, the hairspring is a fine metal wire with a rectangular cross-section that is coiled around itself in the form of an Archimedean spiral. Note that it has a "tail" for the assembly. The hairspring is assembled onto the balance wheel by a collet. Together they form a harmonic oscillator, where the hairspring provides the linear restoring force that reverses the motion of the balance wheel so it oscillates back and forth. As discussed in the previous section, the motion of the balance wheel is approximately a simple harmonic motion, i.e., a sinusoidal motion with a constant period.

The stiffness of the hairspring can be approximately expressed as follows:

$$k_0 = \frac{E \cdot h \cdot t^3}{12L} \tag{4.2}$$

where k_0 is the stiffness of the spring [Nm/rad], E the modulus of elasticity of the spring [GPa], h the height of the hairspring [m], t the thickness [m] and L the length [m].

Along with the moment of inertia of the balance wheel, J_b, the natural frequency of the balance wheel, f_{n0} in s^{-1}, is dependent on the stiffness of the spring and can be approximately determined as follows:

$$f_{n0} = \frac{1}{2\pi}\sqrt{\frac{k_0}{J_b}} \tag{4.3}$$

Note that during the oscillation, the coils of the hairspring shall not touch each other; otherwise, the harmonic oscillation would be jeopardized.

The mainspring, as shown in Fig. 4.2, is a metal ribbon with a rectangular cross-section. It is used to store energy in the form of coiling from winding through turning the knob of the watch or through the automatic winding mechanism. By uncoiling the stored energy drives the movement of the watch. As shown in the figure, the mainspring comes in an assembly, which comprises the barrel

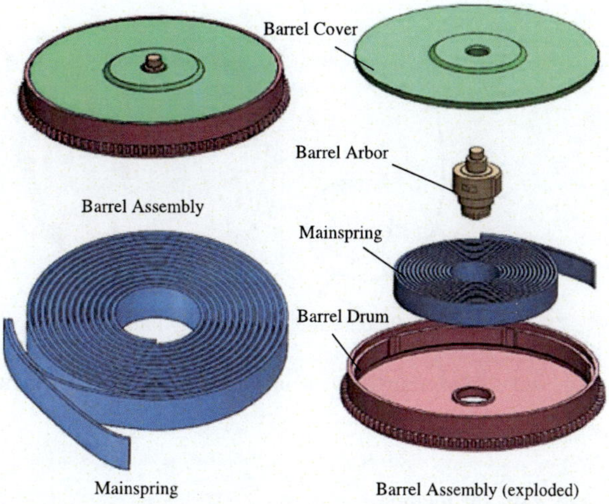

Fig. 4.2 Mainspring and the barrel assembly

drum, the mainspring, the arbor and the barrel cover. The barrel drum houses the mainspring and turns the gear train. The barrel arbor is a pivoted component. One end of the mainspring is attached to the inside of the drum and the other end is assembled onto the arbor, from which the mainspring is wound. The uncoiling of the mainspring drives the drum to output the energy.

Ideally, the barrel drum with the mainspring should provide a constant force for the gear train. Therefore, the mainspring is specially designed as shown in Fig. 4.3a. In its free state, it is in an "S" shape with its outer end coiled in the reverse direction to form an angle exceeding 360°. In the operation, only the inner turns of the mainspring are used, and it is wound around the arbor when fully wounded, as shown in Fig. 4.3b. When fully unwounded, the mainspring is against the outer circumference of the barrel as shown in Fig. 4.3c. The reverse outer turns store extra tension that is available toward the end of the running period. As a result, it provides approximately constant torque during the entire running period of the movement. Figure 4.4 shows the relation between the output force and the unwinding turns obtained from an experiment. From the figure, it is seen that the force output is still not linear, especially when the mainspring is nearly fully unwounded.

The mechanics of the mainspring is rather complex, as it involves changing contacts. Fortunately, thanks to the escapement, the mainspring is not crucial to the timekeeping accuracy of the mechanical watch movement. Thus, we usually neglect the contact force and other factors and use the angular form of Hooke's law to calculate the output torque of the mainspring:

$$\tau = -k\theta + \tau_0 \tag{4.4}$$

where, τ is the torque exerted on the mainspring in [Nm], θ the wound angle of the mainspring in radians and τ_0 the initial load of torque in [Nm].

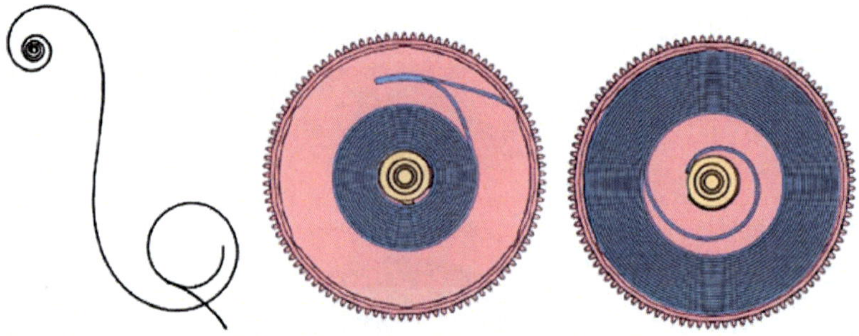

(a) Fully relaxed (b) Fully wound (c) Unwound

Fig. 4.3 Mainspring fully wound (*left*) and fully unwound (*right*). **a** Fully relaxed. **b** Fully wound. **c** Unwound

Fig. 4.4 The relation between the torque and the turns of the mainspring

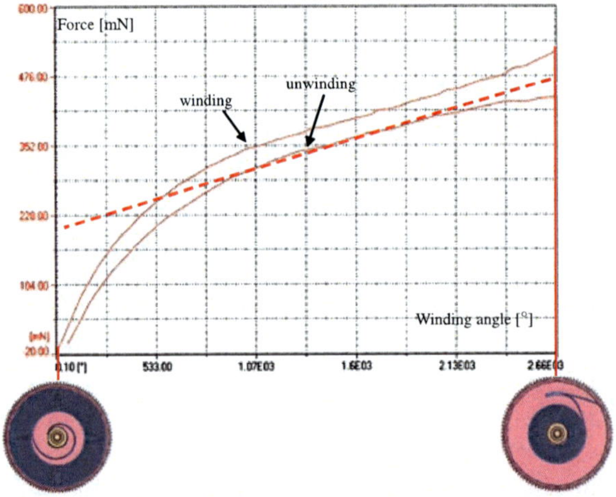

The hairspring is, however, crucial. In fact, it has the highest technical requirement among all the components of the mechanical watch movement because even a small variation may result in a significant effect on the performance of the watch. In this chapter, we focus on the mechanics of the hairspring.

4.2 The Mechanics of the Hairspring

Spring is one of most commonly used mechanical components. Although it seems simple, it can get complex. Throughout the years, much research has been carried out. One of the earliest systematic studies was attributed to Lord Kevin in 1883 (Donkin and Clark 1929). The most commonly used spring is the helical spring. In 1920, because of the demand from automotive engineering, the dynamics of the valve helical spring of the internal combustion engine was investigated in detail (Jehle and Spiller 1929; Love 1944). The other application of the helical spring is found in the vehicle suspension system, and its studies can be dated back to 1940s. In 1966, Wittrick first derived the dynamic model of the helical spring (Lin and Pisano 1987; Wittrick 1966).

In recent years, Lin and Pisano made a significant contribution to the mechanics of the helical spring (Lin and Pisano 1988; Lin and Pisano 1990; Lin et al. 1993). Their study revealed the dynamics of the helical spring in more details, based on which the resonance can be suppressed and the performance can be improved (Cai 2001). Furthermore, it extended the design to variable pitch angle, variable helix radius and different cross-sections with predictable performance (Le and Lin 1994; Tai et al. 1997; Pearson 1982). Now, the dynamics of helical spring is well understood (SAE 1997).

In comparison, spiral spring is much less studied. In the design handbooks, the design formulae for spiral spring are based on much simplified approximation and experiment testing (Spring Manufacturers Institute 2002; Zhang and Liu 1997; Shimoseki et al. 2003). According to the literature, there are only few studies on spiral springs. In Shimoseki et al. (2003) finite element analysis (FEA) was carried on a single turn spiral spring, including both contact and non-contact cases. It shows that the torque-deflection relation for the non-contact case is linear, whereas the contact results in non-linearity. The hairspring, however, has many turns and its mechanics is rather complex.

4.2.1 The Model

As mentioned earlier, the hairspring is Archimedean spiral thin-wall metal wire with rectangular cross section, whose outer endpoint is fixed at the stud pin, ***S***, and the inner endpoint is fixed at the collet shaft, ***C***, which is driven by the balance wheel to exert external torque on the hairspring. Figure 4.5 shows the schematic diagram of hairspring structure. The Cartesian coordinate is set at the center of the collet, and the X axis passes the stud pin.

It is known that for an Archimedean spiral, its governing equation in polar coordinate can be expressed as:

$$r = R_s - a \cdot \theta \tag{4.5}$$

where a is the spiral constant. For the hairspring, the pitch, p, and the number of turns, n, are known. Accordingly, the spiral constant is

$$a = \frac{p}{2\pi} \tag{4.6}$$

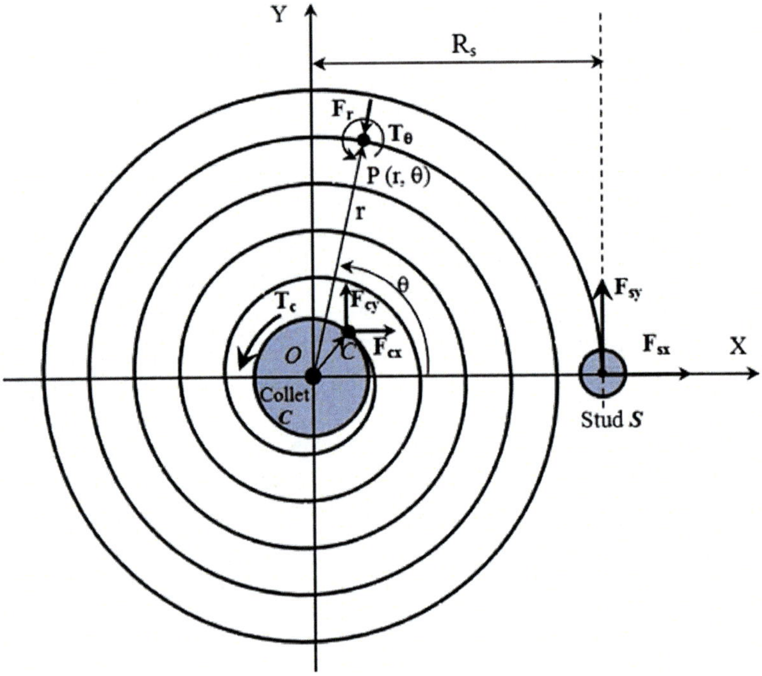

Fig. 4.5 Schematic diagram of the hairspring structure

Thus, the Archimedean spiral spring can be expressed as:

$$r = R_s - \frac{p}{2\pi} \cdot \theta \tag{4.7}$$

The initial boundary conditions are $\theta = \theta_c$ and $r = R_c = R_s - p \cdot n$ at the inner endpoint, i.e., the collet shaft, and $r = R_s$ and $\theta = 0$ at the outer endpoint, i.e., the stud. Any point of the hairspring, $P(r, \theta)$, can be expressed in the Cartesian coordinates, $P(x, y)$, using a simple transform:

$$\begin{cases} x = r \cdot \cos\theta \\ y = r \cdot \sin\theta \end{cases} \tag{4.8}$$

It shall be pointed out that the outer endpoint of the hairspring is fixed, whereas the inner endpoint is turned by the collet shaft from the external torque T_c. There exist two forces, F_{sx} and F_{sy}, at the outer endpoint in X direction and Y direction, respectively; and two forces, F_{cx} and F_{cy}, at the inner endpoint in X direction and Y direction as well. Furthermore, the torque around the origin, as well as the forces in X and Y axis, can be expressed as follows:

$$T_c - F_{cx} \cdot R_c \sin\theta_c + F_{cy} \cdot R_c \cos\theta_c + F_{sy} \cdot R_s = 0 \tag{4.9}$$

$$F_{cx} + F_{sx} = 0 \tag{4.10}$$

$$F_{cy} + F_{sy} = 0 \tag{4.11}$$

where R_c is the radius of the inner endpoint (collet), θ_c is the angle between the inner endpoint and the X axis and R_s the radius of the outer endpoint (stud).

We wish to find the deformation of the hairspring under the external torque. According to the Castigliano's theorem (Wikipedia 2005; Fu 1985), the change of the strain energy under the external torque, $\partial U/\partial T_c$, is equal to the angular displacement of the hairspring–balance wheel, θ_b, that is:

$$\frac{\partial U}{\partial T_c} = \int_L \frac{M}{EI}\frac{\partial M}{\partial T_c} ds = \theta_b \tag{4.12}$$

where U is the strain energy, T_c the external torque (applied onto the collet), M the torque bending moment, I the moment of the inertia and the integration length is the entire hairspring L from the outer endpoint to the inner endpoint.

For any point $P(r,\ \theta)$ of the spring in the polar coordinate, the torque bending moment can be expressed as follows:

$$M(\theta) = F_{cx}(r\sin\theta - R_c\sin\theta_c) - F_{cy}(r\cos\theta - R_c\cos\theta_c) + T_c \tag{4.13}$$

Inserting Eq. (4.13) into Eq. (4.12) and noting that the differential arc of the Archimedean spiral spring is $ds = \sqrt{r^2 + r'^2}d\theta = \sqrt{r^2 + a^2}d\theta$, it follows that

$$\int_0^{\theta_c} \frac{F_{cx}(r\sin\theta - R_c\sin\theta_c) - F_{cy}(r\cos\theta - R_c\cos\theta_c) + T_c}{EI}\sqrt{r^2 + a^2}d\theta = \theta_b \tag{4.14}$$

Accordingly, the deformation of the spring can be found.

The deformation consists of two components: the angular deflection $\alpha(\theta)$ and the radian deflection $\delta(\theta)$. To find the angular deflection $\alpha(\theta)$ of Point $P(r,\ \theta)$, we also use the Castigliano's theorem. Assume an additional torque, T_θ, is being applied, then for any point $Q(r_1,\ \theta_1)$ between point S and point $\boldsymbol{P}$, the bending moment is

$$M(\theta_1) = F_{cx}(r_1\sin\theta_1 - r_c\sin\theta_c) - F_{cy}(r_1\cos\theta_1 - r_c\cos\theta_c) + T_c + T_\theta \tag{4.15}$$

where $\theta_1 \in [0,\ \theta)$ and $r_1 = R_s - a \cdot \theta_1$. On the other hand, for any point $Q(r_2,\ \theta_2)$ between point P to point C, the bending moment is

$$M(\theta_2) = F_{cx}(r_2\sin\theta_2 - r_c\sin\theta_c) - F_{cy}(r_2\cos\theta_2 - r_c\cos\theta_c) + T_c \tag{4.16}$$

where $\theta_2 \in [\theta,\ \theta_c]$ and $r_2 = R_s - a \cdot \theta_2$. Using the Castigliano's theorem, the angular deflection $\alpha(\theta)$ at point $P(r,\ \theta)$ can be expressed as follows:

$$\begin{aligned}\alpha(\theta) &= \frac{\partial U}{\partial T_\theta} = \int_L \frac{M}{EI}\frac{\partial M}{\partial T_\theta}ds \\ &= \int_0^{\theta} \frac{M(\theta_1)}{EI}\frac{\partial M(\theta_1)}{\partial T_\theta}\cdot\sqrt{r_1^2+a^2}\cdot d\theta_1 + \int_\theta^{\theta_c} \frac{M(\theta_2)}{EI}\frac{\partial M(\theta_2)}{\partial T_\theta}\cdot\sqrt{r_2^2+a^2}\cdot d\theta_2\end{aligned} \tag{4.17}$$

Noting that $T_\theta = 0$, and taking derivative of $M(\theta_1)$ and $M(\theta_2)$ with respect to T_θ, and inserting to Eq. (4.17), the angular deflection becomes

$$\alpha(\theta) = \int_0^{\theta} \frac{F_{cx}(r_1\sin\theta_1 - r_c\sin\theta_c) - F_{cy}(r_1\cos\theta_1 - r_c\cos\theta_c) + T_c}{EI}\sqrt{r_1^2+a^2}d\theta_1 \tag{4.18}$$

Similarly, to find the radian deflection $\delta(\theta)$ of point $P(r,\ \theta)$, an additional small force F_r is being applied in the radial direction of $P(r,\ \theta)$; then, for any point $Q(r_1,\ \theta_1)$ between Point S and Point P, the bending moment is

$$\begin{aligned}T(\theta_1) = {} & F_{cx}(r_1\sin\theta_1 - r_c\sin\theta_c) - F_{cy}(r_1\cos\theta_1 - r_c\cos\theta_c) \\ & + T_c + F_r r_1\sin(\theta - \theta_1)\end{aligned} \tag{4.19}$$

where $\theta_1 \in [0,\ \theta)$ and $r_1 = R_s - a\cdot\theta_1$. For any point $Q(r_2,\ \theta_2)$ between point P and Point C, the bending moment is

$$T(\theta_2) = F_{cx}(r_2\sin\theta_2 - r_c\sin\theta_c) - F_{cy}(r_2\cos\theta_2 - r_c\cos\theta_c) + T_c \tag{4.20}$$

where $\theta_2 \in [\theta,\ \theta_c]$ and $r_2 = R_s - a\cdot\theta_2$. Using the Castigliano's theorem, the deflection $\delta(\theta)$ at Point $P(r,\ \theta)$ can be expressed as follows:

$$\begin{aligned}\delta(\theta) &= \frac{\partial U}{\partial F_r} = \int_L \frac{M}{EI}\frac{\partial M}{\partial F_r}ds \\ &= \int_0^{\theta} \frac{T(\theta_1)}{EI}\frac{\partial T(\theta_1)}{\partial F_r}\cdot\sqrt{r_1^2+a^2}\cdot d\theta_1 + \int_\theta^{\theta_c} \frac{T(\theta_2)}{EI}\frac{\partial T(\theta_2)}{\partial F_r}\cdot\sqrt{r_2^2+a^2}\cdot d\theta_2\end{aligned} \tag{4.21}$$

Noting that $F_r = 0$, and taking derivative of $T(\theta_1)$ and $T(\theta_2)$ with respect to F_r, and inserting to Eq. (4.21), the deflection in radial direction becomes

$$\begin{aligned}\delta(\theta) = \int_0^{\theta} & \frac{F_{cx}(r_1\sin\theta_1 - r_c\sin\theta_c) - F_{cy}(r_1\cos\theta_1 - r_c\cos\theta_c) + T_c}{EI} \\ & (r_1\sin(\theta - \theta_1))\sqrt{r_1^2+a^2}d\theta_1\end{aligned} \tag{4.22}$$

In summary, the deformation resulting from the external torque T_c can be obtained from Eqs. (4.18) and (4.22), and the locus of the deformed hairspring can be expressed as

$$\begin{cases} r' = r - \delta(\theta) \\ \theta' = \theta + \alpha(\theta) \end{cases} \tag{4.23}$$

r' and θ' are the radius and angle of the deformed hairspring, respectively.

Because of the deformation, the center of mass of the hairspring, $M(\xi, \eta)$, varies and can be expressed as follows:

$$\begin{cases} \xi = \dfrac{1}{L}\displaystyle\int_0^L x' \cdot \sqrt{1 + (\frac{dy'}{dx'})^2} \cdot dx' \\ \eta = \dfrac{1}{L}\displaystyle\int_0^L y' \cdot \sqrt{1 + (\frac{dy'}{dx'})^2} \cdot dx' \end{cases} \tag{4.24}$$

where $x' = r' \cos\theta'$ and $y' = r' \sin\theta'$; L is the total arc length of the spring, which can be expressed as follows:

$$L = \int_0^{\theta_c} \sqrt{r^2 + a^2} \cdot d\theta \tag{4.25}$$

Assume the spring is placed on the horizontal plane and hence there is no gravitation force, then the strain energy due to bending from the external torque can be expressed as:

$$\begin{aligned} U &= \int_0^{\theta_c} \frac{M^2(\theta)}{2EI} \sqrt{r^2 + a^2} \cdot d\theta \\ &= \int_0^{\theta_c} \frac{[F_{cx}(r\sin\theta - r_c \sin\theta_c) - F_{cy}(r\cos\theta - r_c\cos\theta_c) + T_c]^2}{2EI} \sqrt{r^2 + a^2} \cdot d\theta \end{aligned} \tag{4.26}$$

Furthermore, assume $U = \frac{1}{2} k_{eq} \cdot \theta_b^2$, where k_{eq} is the equivalent stiffness constant of the hairspring, and θ_b is the angle of the balance wheel; then with Eq. (4.26), the stiffness constant is

$$k_{eq} = \frac{\int_0^{\theta_c} [F_{cx}(r\sin\theta - r_c\sin\theta_c) - F_{cy}(r\cos\theta - r_c\cos\theta_c) + T_c]^2 \sqrt{r^2 + a^2} \cdot d\theta}{\theta_b^2 EI} \tag{4.27}$$

The moment of inertia J_s of the hairspring around the collet shaft can be expressed as follows:

$$J_s = \int_0^{\theta_c} \rho \cdot t \cdot h \cdot r'^2 \cdot \sqrt{r'^2 + a^2} d\theta' \tag{4.28}$$

Table 4.1 Parameters of the spring used in the simulation

Parameter	Value
Pitch, p [mm]	0.143
Number of turns, n	13.25
Outer radius, R_s [mm]	2.34
Young's module, E [GPa]	210
Height, H [mm]	0.208
Thickness, T [mm]	0.035
Density, ρ [kg/m^3]	8178
Moment of inertia of balance wheel, J_b [kg·m^2]	3.4×10^{-9}

where ρ, t and h are the density, thickness and height of the spring, respectively.

Furthermore, the natural frequency of the hairspring–balance wheel system after deformation is

$$f_n = \frac{1}{2\pi}\sqrt{\frac{k_{eq}}{J_b + J_s}} \tag{4.29}$$

Because J_s and k_{eq} are the functions of the angular displacement θ_b of the balance wheel, the natural frequency f_n is also the function of θ_b. That is, the natural frequency of the balance wheel and hairspring system varies during the rotation of the balance wheel. Therefore, the average natural frequency for the balance wheel and hairspring system is used, which can be calculated by the following equation.

$$\bar{f}_n = \frac{1}{2\alpha}\int_{-\psi}^{\psi} f_n(\theta_b) \cdot d\theta_b \tag{4.30}$$

where ψ is the rotation amplitude of the balance wheel.

As mentioned earlier, the frequency of the balance wheel–hairspring system, f_{n0}, can be approximated by Eq. (4.3). Now considering the influence of the deformation and the moment of inertia of the hairspring, the actual frequency becomes $\bar{f}_n$ and the variation can be expressed as follows:

$$\Delta f_n = \frac{\left|\bar{f}_n - f_{n0}\right|}{f_{n0}} \tag{4.31}$$

4.2.2 Computer Simulation

Using the model derived above, computer simulation is conducted with MATLAB®. In the simulation, the spring is discretized into a number of sections and each section is 5° apart. The external torque is assumed to be 1.07×10^{-8} kg m^2. The key parameters of the spring are summarized in Table 4.1.

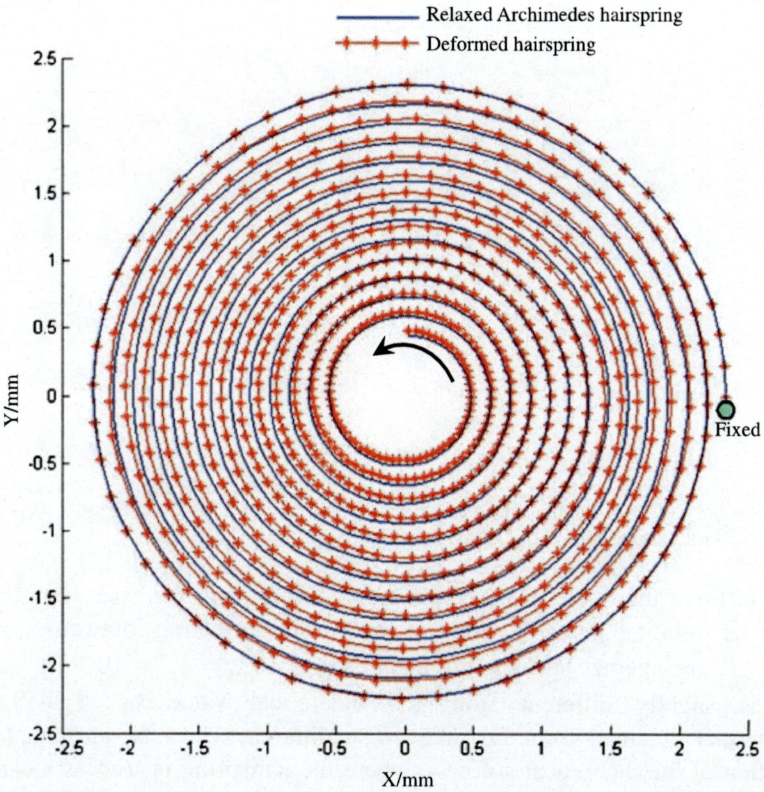

Fig. 4.6 The hairspring in its rest state (solid line) and its deformed position after it is rotated 360° counterclockwise

Figure 4.6 shows the spring before and after being rotated 360° counterclockwise. *The solid line* is the location of the spring in the rest state, whereas *the red line* is the location of the spring after the rotation and *the red stars* denote the position of the discretized sections. From the figure, it is seen that the deformation is not uniform. It gradually increases from the fixed outer endpoint, reaches the maximum in the middle and then decreases towards the inner end point. Note that one extra turn is being added because of the rotation. The clockwise deformation is symmetric. Furthermore, their combination generates a wave motion.

Because of the deformation, the center of the mass of the spring varies. Figure 4.7 shows the locus of the center of the mass in both the polar coordinate and the Cartesian coordinate. From the figure, it is seen that the center of mass is never in the rotation center (it is about 1.4 μm away from the rotation center). It moves close to the rotation center when rotating counterclockwise (inwards rotation) and moves away when rotating clockwise (outwards rotation). The variation is about 0.2 μm. Clearly, this will affect the timekeeping accuracy, as discussed in the subsequent section.

Because of the variation of the mass center, the equivalent stiffness of the hairspring with respect to the balance wheel rotation varies as well, as shown in

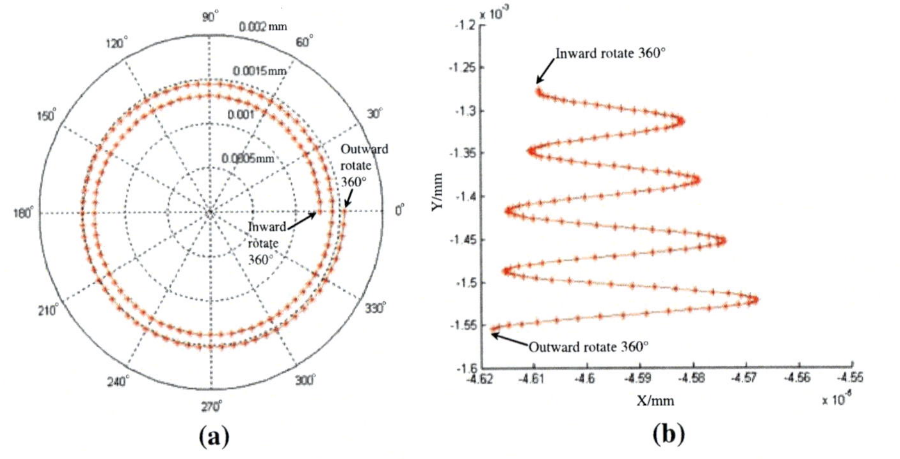

Fig. 4.7 Locus of the center of mass of the hairspring when rotating from $-360°$ to $360°$. **a** In the polar coordinate system **b** In Cartesian coordinate system

Fig. 4.8. From the figure, it is seen that the stiffness will increase when the hairspring is rotating inwards and decrease when rotating outwards. pplying Eq. (4.27), the average stiffness of the hairspring is $k_{eq} = 1.7 \times 10^{-6}$ $(N \cdot m/rad)$. This is slightly different from the theoretical value $k_0 = 1.3459 \times 10^{-6}$ $(N \cdot m/rad)$ obtained from Eq. (4.2). The difference may be attributed to the definition of the theoretical stiffness, where the hairspring is seen as a cantilever beam expanded from the Archimedes curve and its length is unchanged during rotation. Actually, the hairspring is only initially in Archimedes form, it is deformed due to the rotation, which results in the change of the hairspring stiffness.

Because of the deformation of the hairspring, its moment of inertia around the collet shaft varies as well, as shown in Fig. 4.9. From the figure, it is seen that the moment of inertia is dependent on the rotation of the balance wheel. Fortunately, it is very small comparing to that of the balance wheel; thus, its influence on the natural frequency is small.

Figure 4.10 shows the natural frequency of the hairspring–balance wheel system with respect to the balance wheel rotation. It is seen that the hairspring deformation will influence the natural frequency. Specifically, the average natural frequency is 3.5471 (1/s), which is 12% in difference comparing to the frequency of 3.1665 (1/s) calculated using Eq. (4.3). In addition, from the data in Fig. 4.10, the variation of natural frequency is about 2.6%.

4.3 The Effects of the Hairspring and the Tourbillon

In Chap. 3, the Swiss Lever Escapement was studied under the approximation of negligible hairspring mass, and the resulting movement was described by a simple harmonic oscillator driven by the escape wheel through the pellet fork. From the

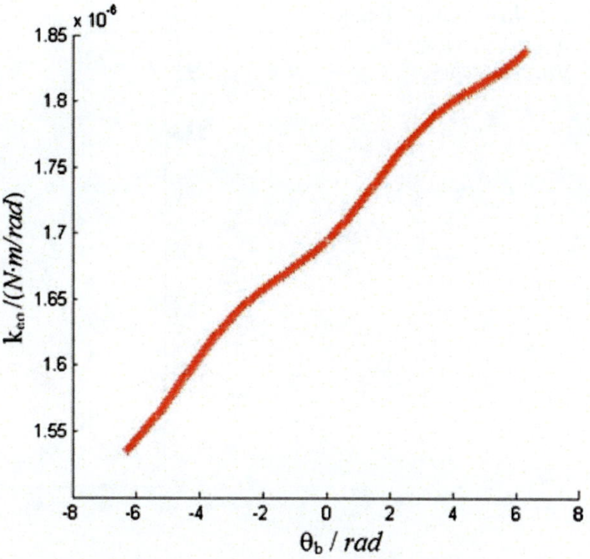

Fig. 4.8 Equivalent stiffness of the hairspring with respect to the balance wheel rotation

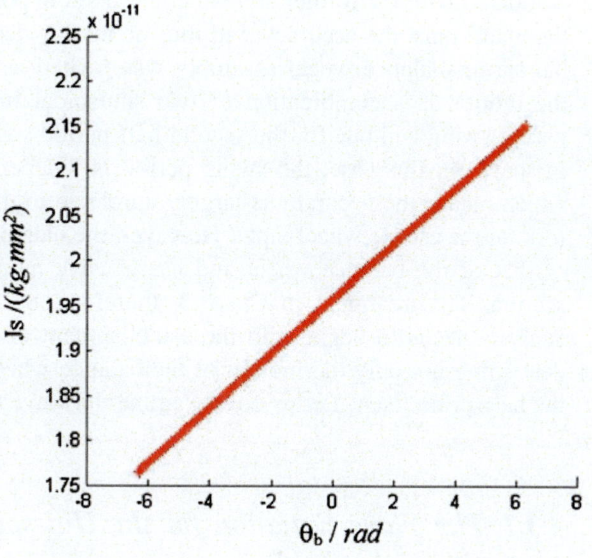

Fig. 4.9 Moment of inertia of the hairspring with respect to the balance wheel rotation

previous section, however, it is seen that the motion of the hairspring is not a simple harmonic. To understand its influence, a closer examination of the dynamics of the hairspring–balance wheel assembly is warranted. This becomes even more interesting when a much sophisticated "Tourbillon" mechanism—regarded as the most fascinating miniature handmade mechanism—is involved Xu et al. (2011).

As shown in Chap. 3, the timekeeping accuracy of the Swiss Lever Escapement is largely determined by two factors, the swing of the balance wheel and the impulsive push of the pallet fork. Because the latter is much shorter in duration

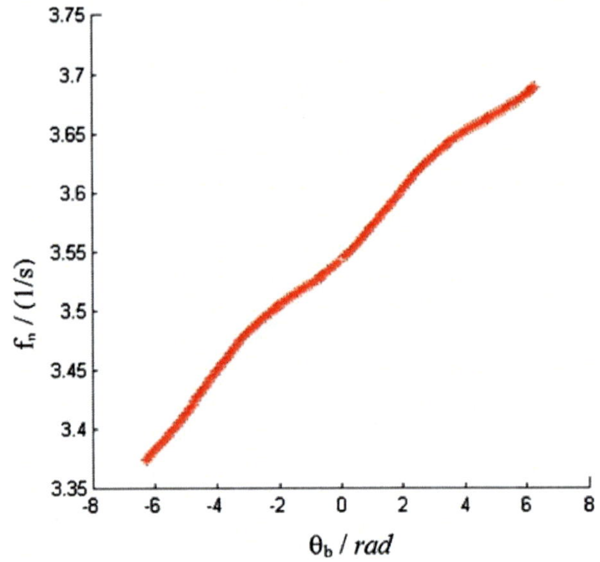

Fig. 4.10 Natural frequency of the hairspring-balance wheel system

compared with the former and is largely fixed in position (angular coordinates) by the guard pins, the accuracy will thus be mainly decided by the former. When the hairspring–balance wheel assembly was treated as a simple harmonic oscillator, the motion is a combination of two sinusoidal functions (one for the first-half period swing and one for the second-half period swing) and thus becomes circular in shape. In this case, the swing period is independent on the swing amplitude, which means the accuracy is largely unrelated to the impulse pushes of the pallet fork, or the escape wheel input. However, the hairspring usually weighs about 10% of that of the balance wheel and has a wave motion, as shown in the previous section. The treatment in Chap. 3, therefore, becomes inadequate. The correct analysis needs to begin with the establishment of a partial differential equation, describing not only the motion of the balance wheel but also the wave motion of the hairspring itself. Let us now examine the wave equation of the hairspring first.

4.3.1 The Wave Equation for the Hairspring–Balance Wheel Assembly

(1) Wave equation, boundary conditions and eigenvalues

First, we set the coordinate as shown in Fig. 4.11 and denote

$M =$ the mass of the balance wheel ($\approx$85 milligrams from a catalog of Swatch^{TM}),
$m =$ the mass of the hairspring ($\approx$12 milligrams; Swatch^{TM}),
$\ell =$ the radian length of the hairspring,
$S =$ the cross section of the hairspring,

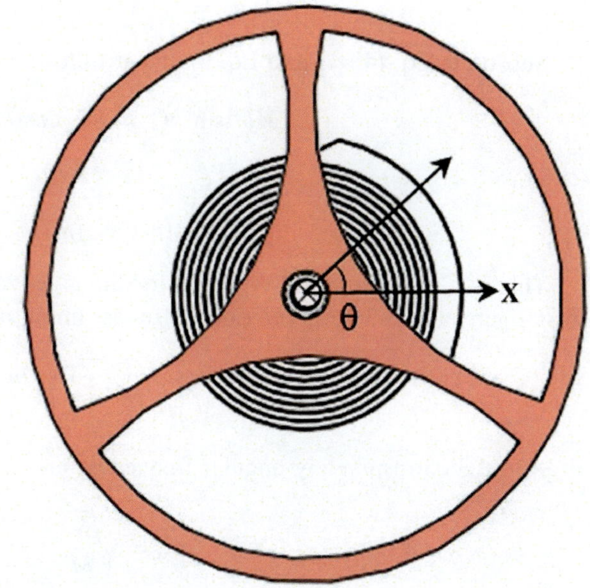

Fig. 4.11 Illustration of the hairspring—balance wheel assembly

E = Young's modulus of the "linear spring,"
x = the radian coordinate along the spring length,
$u(x, t)$ = the displacement of the "linear spring" parts from their equilibrium positions;

For a small part of the mass of the linear spring near x, $u_x = \partial u/\partial x$ = strain, and the equation of motion is:

$$\frac{m}{\ell}\mathrm{d}x\,u_{tt} = ES[u_x(x+\mathrm{d}x) - u_x(x)] = ES\,u_{xx}\,\mathrm{d}x \tag{4.32}$$

where the subscripts, x and t, denote the partial derivatives against x and t, respectively. Equation (4.32) can be rewritten as:

$$u_{tt} - -a^2 u_{xx} = 0 \tag{4.33}$$

where $a^2 = E/(m/S\ell)$, which is equal to E/(density of the spring). This is a one-dimensional wave equation. For the fixed boundary condition $u(x = 0) = 0$, its solution is:

$$u(x,t) = \sum_n u_n \sin(k_n x)\exp(i\omega_n t) \tag{4.34}$$

where $\omega = a \cdot k$, and k is the wave number. Note that only the real part is useful. To determine the wave number, the boundary condition at $x = \ell$ is needed, which is neither the first, second or third type. Rather, it is a modified Cauchy type, where the acceleration of the mass, the balance wheel, is governed by the tension at the tail of the spring:

$$M\,u_{tt} = -ES\,u_x\ (at\ x = \ell) \tag{4.35}$$

Substitute Eq. (4.34) into Eq. (4.35), it follows that:

$$M\omega^2 \sin(k\ell) = ES\,k\cos(k\ell)$$

or

$$(k\ell)\tan(k\ell) = (m/M) \tag{4.36}$$

This is a transcendental equation for the eigenvalues, K_n, of the system. The first eigenvalue can be approximated by assuming $m \ll M$, which results in:

$$k_0\ell \approx \sqrt{\frac{m}{M}}\left(1 - \left(\frac{1}{6}\right)\left(\frac{m}{M}\right)\right) \tag{4.37}$$

And the corresponding angular frequency, ω_0, is

$$\omega_0 = \sqrt{\frac{K}{M}} \tag{4.38}$$

where $K = ES/\ell =$ the "linear spring constant." This is exactly the case when ignoring the mass of the hairspring, m, and the whole spring is linearly elongated in unison.

However, as shown in Eq. (4.34), the overall oscillation is not simple harmonic, because there are high-order harmonics for $n = 1, 2, 3, \ldots$

$$k_n\ell \approx n\pi + \frac{m/M}{n\varpi} \tag{4.39}$$

and

$$\omega_n = a \cdot k_n \tag{4.40}$$

Furthermore, the ratio of the first-order harmonics versus the base frequency is roughly given by:

$$\frac{\pi}{\sqrt{m/M}} \approx 10\ (\text{when } m/M \approx 1/8 - 1/9)$$

In other words, the higher-order harmonics will contribute approximately 10% of the total oscillations.

(2) Eigen functions and Fourier series by non-orthogonal bases

Due to the complex boundary condition at $x = \ell$, the corresponding eigenfunctions may not form an orthogonal set. Instead, it can be shown, using the boundary conditions, that

$$\int_0^\ell \sin(k_n x)\sin(k_j x)\,\mathrm{d}x + \ell\frac{M}{m}\sin(k_n\ell)\sin(k_j\ell) = 0,\ j \neq n \tag{4.41}$$

and

$$\int_0^{\ell} \sin(k_n x)\sin(k_n x)\mathrm{d}x + \frac{\ell}{2}\frac{M}{m}\sin(k_n\ell)\sin(k_n\ell) = \frac{\ell}{2},\ j = n \tag{4.42}$$

Therefore, the initial conditions, either $u(x, t = 0)$ or $u_t(x, t = 0)$, can still be incorporated to settle the Fourier coefficients u_n. The latter can be complex numbers, with the real part corresponding to the initial displacement, and the imaginary part to the initial velocity. For example, integrating Eq. (4.34) over x with $\sin(k_m x)dx$, it follows that

$$\int_0^{\ell} \sin(k_j x)u(x,0)\mathrm{d}x = \sum_n \mathrm{Re}(u_n)\int_0^{\ell} \sin(k_j x)\sin(k_n x)\mathrm{d}x \tag{4.43}$$

Because

$$u(\ell,0) = \sum_n \mathrm{Re}(u_n)\sin(k_n l)$$

we may construct an additive term to nullify the non-orthogonal terms,

$$u(\ell,0)\ell\frac{M}{m}\sin(k_m\ell) = \sum_n \mathrm{Re}(u_n)\ell\frac{M}{m}\sin(k_n\ell)\sin(k_m\ell) \tag{4.44}$$

by summing both sides of Eqs. (4.43) and (4.44), and utilizing Eqs. (4.41) and (4.42), it follows that

$$\int_0^{\ell} \sin(k_j, x)u(x,0)\mathrm{d}x + u(\ell,0)\ell\frac{M}{m}\sin(k_j\ell) = \mathrm{Re}(u)\frac{\ell}{2}\left(1+\frac{M}{m}\right)\sin^2(k_m\ell)$$

To resolve for $\mathrm{Re}(u_j)$:

$$\mathrm{Re}(u_j) = \frac{\int_0^{\ell} \sin(k_j x)u(x,0)\mathrm{d}x + u(\ell,0)\ell\frac{M}{m}\sin(k_j\ell)}{\frac{\ell}{2}\left[1+\frac{M}{m}\sin^2(k_j\ell)\right]} \tag{4.45}$$

Note that the denominator can also be rewritten as

$$\frac{\ell}{2}\left(1+\frac{M}{m}\sin^2(k_j\ell)\right) = \frac{\ell}{2}\left(1+\frac{mM}{m^2+M^2(k_j\ell)^2}\right) \tag{4.45a}$$

Similarly, we can solve for $\mathrm{Im}(u_j)$ using the same strategy:

$$\mathrm{Im}(u_j) = \frac{\int_0^{\ell} \sin(k_j x)u_t(x,0)\mathrm{d}x + u_t(\ell,0)\ell\frac{M}{m}\sin(k_j\ell)}{ak_m\frac{\ell}{2}\left[1+\frac{M}{m}\sin^2(k_j\ell)\right]} \tag{4.46}$$

(3) Relative amplitude of the high order harmonics

The amplitudes of various orders of harmonics may be analyzed by the denominators of Eqs. (4.45) and (4.46). It can be argued as follows: In the middle of the swing, the initial conditions for a half-period $u(x, 0)$ and $u_t(x, 0)$ can be rather arbitrary, not only because the torque input from the pallet fork but also because the initial condition of the hairspring position. This generates a quasi-periodic motion as shown by the simulations presented in Chap. 3. It is similar to the ergodic hypothesis in a system with a large number of degree-of-freedom (DOF). Thus, we could only find a relative ratio. The first high-order harmonics versus the base frequency oscillation is

$$\frac{1+\frac{M}{m}\sin^2(k_0\ell)}{1+\frac{M}{m}\sin^2(k_1\ell)} \approx 2 = O\left(10^0\right) \tag{4.47}$$

in terms of the initial positions, and

$$\frac{k_0}{k_1}\frac{1+\frac{M}{m}\sin^2(k_0\ell)}{1+\frac{M}{m}\sin^2(k_1\ell)} \approx 0.2 = O\left(10^{-1}\right) \tag{4.48}$$

in terms of the initial velocities.

When more specific knowledge of the initial conditions is available, the analysis may be extended to the numerators to yield a more explicit picture. For example, one of the special cases is when $u(x, 0) = u_t(x, 0) = 0$, except $u_t(\ell, 0)$. The numerators of $\mathrm{Im}(u)$ can then be evaluated as:

$$\int_0^\ell \sin\left(k_j, x\right)u_t(x,0)\mathrm{d}x + u_t(\ell,0)\ell\frac{M}{m}\sin\left(k_j\ell\right) = u_t(\ell,0)\frac{\ell M}{m}\sin\left(k_j\ell\right)$$

Therefore, the wave amplitude ratio of the first-order harmonics versus that of the base frequency becomes

$$\frac{\mathrm{Im}(u_1)}{\mathrm{Im}(u_0)} \approx \frac{0.2\sin(k_1\ell)}{\sin(k_0\ell)} \approx 0.2\frac{\sqrt{m/M}}{\pi} \approx 0.02$$

In addition, because the amplitude of the oscillating mass (i.e., the swing of the balance wheel) is governed by the eigen functions evaluated at $x = \ell$, another factor $\sin(k_1\ell)/\sin(k_0\ell)$ needs to be multiplied onto the above expression. Therefore, the resulting amplitude ratio between the first-order harmonics and that of the base frequency is

$$0.02\frac{\sin^2(k_2\ell)}{\sin^2(k_1\ell)} = O\left(10^{-3}\right) \tag{4.49}$$

This implies that the high-order harmonics will result in an error in the order of 10^{-3}. Such a seemingly small error is equivalent to an error of $10^{-3} \times 60$ (s/min) $\times$ 60 (min/h) $\times$ 24 (h/day) = 86.4 s, or more than one minute per day. Thus, it cannot be dropped in the detailed analysis of the escapement. The immediate consequence of this is the phase trajectory $[u(\ell), u_t(\ell)]$ can no longer be described simply by circles but has to be modified to include high-order oscillations. Due to the fact that the frequencies of the high-order harmonics are not commensurate with that of the base, nor are they commensurate with each other, plus the fact that the input torque cannot be kept constant, the various orders of the harmonics will sooner or later be out of phase against each other. In other words, the initial conditions will be different for the subsequent swings, which lead to a different combination of various orders. Therefore, the resulting trajectory will be changing with time. In other words, the mechanical watch would have an error in the order defined in Eq. (4.49).

4.3.2 More Precise Description of the Hairspring Movement by Fourth-Order Differentials

(1) From a bending beam to the coiled hairspring

The mechanics of a bending beam is well known: For an elastic slab of cross-section b (=width) and c (=thickness, along the bending direction), the amount of deformation at the free end $u(\ell)$ (parallel to c) is given by

$$u(\ell) = V(\ell)\frac{\ell^3}{3EI} \tag{4.50}$$

where V is the loading force at the end of the beam and I is the cross-section area, obtainable by

$$I = b\int_c y^2 dy = \frac{bc^3}{12} \tag{4.51}$$

Locally, in the vicinity of x along the beam, the bending produces a curvature, which can be approximated by u_{xx} (ignoring $\left(1+u_x^2\right)^{3/2}$ as the denominator). It is related to the local bending torque $T(x)$ through the lateral elastic deformation across the thickness c, part of which (above the neutral plane) is elongated and the other (below the neutral plane) is compressed, which gives rise to the same area moment:

$$T(x) = EIu_{xx} \tag{4.52}$$

Now, let us consider the spiral spring as the coiled bending beam. Eq. (4.52) acts as a pivotal point; on one side, the integral of the curvature leads to some expansion of the spring tail, because the angular displacement, Θ, is given by arc/radius, or $dx \times$ curvature, i.e.,

$$\Theta(\text{the balance wheel swing}) = \int_0^\ell u_{xx}\mathrm{d}x = u_x(\ell) \tag{4.53}$$

On the other hand, it is related to the shear force, experienced by the cross-section of the beam, $V(x)$, through the balance of the torques within x and $(x + dx)$:

$$V(x) = -\frac{\mathrm{d}T(x)}{dx} = -EIu_{xxx} \tag{4.54}$$

This can be further differentiated to obtain the net force applied to the mass, via the balance of the forces within x and $(x + dx)$:

$$\text{net force} = \frac{\mathrm{d}V(x)}{dx} = -EIu_{xxxx} \tag{4.55}$$

To get a quantitative picture, let us consider a static beam fixed at $x = 0$ and supplied with a load of $V(\ell)$ at $x = \ell$. Neglect the gravity of the beam itself, we have

$$\begin{aligned}
&\text{net force} = \tfrac{\mathrm{d}V(x)}{dx} = -EIu_{xxxx} = 0 \text{ (except at both ends)};\\
&\text{shear force} = V(x) = -\frac{dT(x)}{dx} = -EIu_{xxx} = \text{constant} = V(\ell); \qquad (4.56)\\
&\text{torque} = T(x) = EI\,u_{xx} = V(\ell)\,(\ell - x) = \text{linear function of } x;\\
&\text{expansion} = u_x(\ell) = \int_0^\ell u_{xx}\mathrm{d}x = V(\ell)\frac{(\ell x - x^2/2)}{EI} = \text{quadratic form of } x;
\end{aligned}$$

And finally, the displacement is

$$u(x) = V(\ell)\frac{\left(\frac{\ell x^2}{2} - \frac{x^3}{6}\right)}{EI}$$

This leads to Eq. (4.50) at $x = \ell$.

(2) Wave equation of the bending coil spring and the boundary condition
From Eq. (4.50) it is obvious that the motion of the coil spring at x follows:

$$\frac{m}{\ell}u_{tt} = EIu_{xxxx} \tag{4.57}$$

or

$$u_{tt} + (a')^2 u_{xxxx} = 0, \text{ where } (a')^2 = \frac{EI}{m/\ell}$$

This is a fourth-order partial differential equation. To find the solution, we need four boundary conditions, with two on each end of the spring. Assuming the spring is fixed at $x = 0$, examining the last part of the previous section, we get:

$$u(x = 0) = 0 \text{ and } u_x(x = 0) = 0 \tag{4.58}$$

Note that higher order derivatives of u (against x) are not vanishing at $x = 0$. (If the end is free, i.e., $T = 0$, then both $u_{xx} = u_{xxx} = 0$; but if the end is supported, then $u = u_{xx} = 0$.) For the boundary of $x = \ell$, however, things are a little different, because both u and u_x are non-zero and unknown. Again by examining the last part of the previous section, we find that u_{xx}, which is related to the local torque, vanishes. In addition, because the balance wheel is connected to the spring tail, similar to Eq. (4.58), we will have:

$$u_{xx}(\ell) = 0 \text{ and } I_M\Theta_{tt} = I_M u_{xtt} = -\ell V(\ell) = \ell EI\, u_{xxx} \tag{4.59}$$

where I_M is the angular inertia of the balance wheel (its mass M times R^2, where R is usually about 3~5 mm), and the extra factor ℓ is obtained by the following consideration: The virtual work performed by the shear force at the tail is given by $V(\ell) \times \delta u(\ell)$, which is equivalent to that by the torque on the balance wheel (I_M) times $\delta\Theta$, or $\delta u_x(\ell)$.

(3) Solution to the coil spring wave equation and the eigenvalues

The usual strategy of variable separation is employed to solve Eq. (4.57), where $u(x, t)$ is composed of the time part (satisfying a second-order oscillatory equation), multiplied by the coordinate part (governed by a fourth-order linear ordinary differential equation), which can be written as (again taking only the real part):

$$u(x,t) \sim [A\,\mathrm{ch}(kx) + B\,\mathrm{sh}(kx) + C\cos(kx) + D\sin(kx)]\exp(i\omega t) \tag{4.60}$$

where A, B, C and D are constants. Incorporating boundary conditions at $x = 0$ of Eq. (4.58):

$$u(x) = A[\mathrm{ch}(kx) - \cos(kx)] + B[\mathrm{sh}(kx) - \sin(kx)] \tag{4.61}$$

To determine the eigenvalues of k, we need to use the boundary condition of $x = \ell$. The first half of Eq. (4.57), $u_{xx}(\ell) = 0$, yields:

$$A[\mathrm{ch}(k\ell) + \cos(k\ell)] + B[\mathrm{sh}(k\ell) + \sin(k\ell)] = 0 \tag{4.62}$$

And labeling $r = \frac{m}{M}\left(\frac{1}{Rk}\right)^2$, the second half gives:

$$A[(1+r)\mathrm{sh}(k\ell) + (1-r)\sin(k\ell)] + B[(1+r)\mathrm{ch}(k\ell) - (1-r)\cos(k\ell)] = 0 \tag{4.63}$$

To obtain the non-vanishing A and B, the determinant of their coefficients in Eqs. (4.62) and (4.63) must be zero, i.e.,

$$\mathrm{sh}(k\ell)\sin(k\ell) = r\,[1 + \mathrm{ch}(k\ell)\cos(k\ell)] \tag{4.64}$$

which is again a transcendental equation for the eigenvalues of Eq. (4.57).

Although the direct solution to Eq. (4.64) is not available, we may qualitatively argue for the asymptotic values of k. For example, when $m \ll M$ (mass of the hairspring $\ll$ mass of the balance wheel), $(k\ell) \ll 1$. Or, $\mathrm{sh}(k\ell) \approx \sin(k\ell) \approx (k\ell)$, and $\mathrm{ch}(k\ell) \approx \cos(k\ell) \approx 1$, we get:

$$(k_0\ell)^4 \approx 2r = 2\frac{m}{M}\left(\frac{\ell}{R}\right)^2 \text{ or } (k_0\ell)^2 \approx \sqrt{\frac{2m}{M}}\left(\frac{\ell}{R}\right) \tag{4.65}$$

The corresponding "base frequency," ω_0, is:

$$\omega_0 = a'(k_0)^2 = \sqrt{\frac{2EI}{\ell R^2 M}} = \sqrt{\frac{K'}{M}} \tag{4.66}$$

where K' is the equivalent spring constant, given by $\frac{2EI}{\ell R^2}$, (cf. Eq. 4.4). Comparing Eqs. (4.65) and (4.36), it is seen that the "base frequency" here is also manifested by a "unison" spring attached to the connected mass of the balance wheel. On the other extreme, when $(k\ell) \gg 1$, $[\mathrm{sh}(k\ell)\sin(k\ell)]^{-1} \approx 0$, we get:

$$r \approx \tan(k\ell), \text{ or } \frac{m}{M}\left(\frac{\ell}{R}\right)^2 \approx (k\ell)^2 \tan(k\ell)$$

thus,

$$k\ell \approx n\pi \tag{4.67}$$

which is also similar to Eq. (4.37), except, there is extra factor of (ℓ/R). The latter is a little tricky because, ℓ, the total length of the hairspring can be quite a few multiples of R.

Beyond the eigenvalues, parallel discussions in Sect. 4.3.1 can be considered, although they may not be expressed in terms of analytical forms. For example, the same eigen function integrals as Eqs. (4.38) and (4.39) could be constructed, which may be utilized to incorporate the initial conditions into Fourier series, to be followed by the amplitude analysis. Similar results are expected, because the solutions again involving incommensurate "harmonics."

4.3.3 The Case of Tourbillon

Our analysis did not involve gravity so far, which would certainly interfere the resulting motion, unless the watch is facing up or down (dial up or dial down) when gravity is perpendicular to the motion of the spring assembly. To counteract this, a pricy mechanism called Tourbillon has been developed by watchmakers. It is accomplished by mounting the escapement in a rotating frame, so that the effect of gravity cancels out when the escapement is rotated to the opposite position. It usually coincides with the minute wheel, so that a circle is completed every 60 seconds. Without getting into its charming sensation and fascinating craftsmanship, let us begin the analysis by adding a gravity force as a source into Eq. (4.33):

$$u_{tt} - a^2 u_{xx} = f(x \cdot t), \quad \left(f = \frac{F}{m/\ell} = \text{reduced forces}\right) \tag{4.68}$$

Depending on the orientation, the action of gravity alone can be approximated by $f(x) \sim \sin(x/R)$, where R is the average radius of the spring coil. The effect of Tourbillon mechanism, therefore, is displayed by the modification of $\sin(x/R)$ to become:

$$u_{tt} - a^2 u_{xx} = g \sin\left(\frac{x}{R} - \frac{2\pi t}{\tau}\right) \tag{4.69}$$

where $\tau = 60$ seconds and g is the gravity constant. It should be pointed out that we have kept the discussion simple by using the second-order equation. Also the vertical orientation is assumed, otherwise (e.g., crown up, down or left) there will be an extra cosine factor to reduce the effect of g. Furthermore, other forces such as Coriolis' are deemed insignificant and dropped.

The solution to Eq. (4.69) can be found by two separate parts, one to accommodate the "source," $f(x, t)$, and the other the initial conditions as before. However, both parts should meet the same boundary conditions outlined near Eq. (4.34). In other words, set $u = u' + u''$. While u'' is obtained from Sect. 4.3.1 to account for the initial conditions $u(x, 0)$ and $u_t(x, 0)$ but with zero source (homogeneous equation), while u' is resolved from Eq. (4.69) but under zero initial values. Furthermore, u' should be constructed by the eigen-function set found in Eq. (4.34), to comply the same boundary conditions. To solve for u', we may utilize the trick of converting a heterogeneous equation (with zero initial values) into a homogeneous one but with non-zero initial velocity. It is based on the fact that, because the equation is linear, the source $f(x, t)$ can be divided into infinite number of delta functions (pulses) located at t', each of which may then be treated as an initial velocity at t'. The solution to each delta source, $v(x, t)$, can then be integrated to form the solution to the heterogeneous equation, $u'(x, t)$:

$$v_{tt} - a^2 v_{xx} = 0; \quad v(x = 0) = 0; \quad M v_{tt} = -ES\, v_x \ (x = \ell) \tag{4.70}$$

$$v\left(x,\ t = t' + 0\right) = 0; \quad v_t\left(x,\ t = t' + 0\right) = g \sin\left(\frac{x}{R} - \frac{2\pi t}{\tau}\right) \tag{4.71}$$

where t' is first regarded as a *fixed* parameter, and $u'(x, t)$ becomes

$$u'(x,\ t) = \int_0^t v(x,\ t,\ t')\, \mathrm{d}t' \tag{4.72}$$

The solution to Eqs. (4.70) and (4.71), following Sect. 4.3.2, is

$$v(x,\ t,\ t') = \sum_n v_n \sin(k_n x) \exp[i\omega_n(t - t')] \tag{4.73}$$

Because the initial displacement of $v = 0$, only the imaginary part of v_n needs to be evaluated, which becomes (cf. Eq. 4.44):

$$\mathrm{Im}(v_j) = \frac{\int_0^\ell \sin\left(k_j x\right) v_t(x,\ t')\mathrm{d}x + v_t(\ell,\ t')\frac{\ell M}{m}\sin\left(k_j \ell\right)}{\frac{a k_{m\ell}}{2}\left(1 + \frac{M}{m}\sin^2\left(k_j \ell\right)\right)} \tag{4.74}$$

Plug Eq. (4.71) into Eq. (4.74), the numerators become (divided by g):

$$\int_0^{\ell} \sin(k_j x)\sin\left(\frac{x}{R}-\frac{2\pi t'}{\tau}\right)\mathrm{d}x+\sin\left(\frac{\ell}{R}-\frac{2\pi t'}{\tau}\right)\frac{\ell M}{m}\sin(k_j\ell)$$

Therefore, the final form of $u'(x,\,t)=\sum_n u'_n(t)\sin(k_n x)$ and $(\alpha = 2\pi\,/\,\tau)$:

$$u'_n(t)=\frac{g\left[\int_0^{\ell}\mathrm{d}x\sin(k_0 x)\int_0^{t}\mathrm{d}t'\sin\left(\frac{x}{R}-\frac{2\pi t'}{\tau}\right)\sin(\omega_n t-\omega_n t')+\frac{\ell M}{m}\sin(k_j\ell)\int_0^{t}\mathrm{d}t'\sin\left(\frac{\ell}{R}-\frac{2\pi t'}{\tau}\right)\sin(\omega_n t-\omega_n t')\right]}{\frac{ak_{m\ell}}{2}\left(1+\frac{M}{m}\sin^2(k_j\ell)\right)} \tag{4.75}$$

The numerator of Eq. (4.75) may be further computed to become:

$$g\int_0^{\ell}\mathrm{d}x\sin(k_n x)\frac{\omega_n}{\omega_n^2-\alpha^2}\sin\left(\frac{x}{R}-\alpha t\right)-\frac{\sin\left(\frac{x}{R}-\omega_n t\right)}{2\omega_n-2\alpha}-\frac{\sin\left(\frac{x}{R}+\omega_n t\right)}{2\omega_n+2\alpha}$$

$$+g\frac{}{m}\sin(k_j\ell)\frac{\omega_n}{\omega_n^2-\alpha^2}\sin\left(\frac{\ell}{R}-\alpha t\right)-\frac{\sin\left(\frac{\ell}{R}-\omega_n t\right)}{2\omega_n-2\alpha}-\frac{\sin\left(\frac{\ell}{R}+\omega_n t\right)}{2\omega_n+2\alpha}$$

Keep in mind that these are added on top of $u''(x,\,t)$ due to the original free oscillation as shown in Sect. 4.3.1, in no way can they be cancelled by the gravity part $u'(x,\,t)$. Therefore, the overall motion in a Tourbillon watch becomes much more complex, instead of being simplified. Without further calculation, it is obvious that the $u'(x,\,t)$ is entangled by both the eigenvalues, ω_n, and the per minute rotation, α. They are in competing magnitudes, because three pre-factors in the second line are all proportional to ω_n^{-1}. Within the content of u', we can reach the same order of magnitude of the ratio between the first harmonics and the base frequency, because the main change against k_j is found in the second line, which is again depending on $\sin(k_j\ell)$, similar to that in Eq. (4.46). In short, the Tourbillon is not necessary good for the precision timekeeping.

4.4 Concluding Remarks

In this chapter, we studied the hairspring and the hairspring–balance wheel assembly. The hairspring is a crucial component of the mechanical watch movement. The simple formula found in the existing literature is subject to an error of 10%, as it does not count the wave motion of the hairspring, which causes the change of the mass center as well as the stiffness and hence, the oscillating frequency. The wave motion of the hairspring has a significant effect on the hairspring–balance wheel assembly. Though, the effect is non-linear and hence, difficult to compensate. This is why the spring–balance wheel assembly still greatly depends on manual tuning by skilled craftsman. The Tourbillon adds even more complex motion to the hairspring–balance assembly. In fact, it probably causes more inaccuracy then compensates the gravity force.

References

Fu BL (1985) On the modified Castigliano's theorem. Appl Math Mech 5(2):1263–1272. doi:10.1007/BF01895122

Cai X (2001) Dynamic modeling of helical springs with design parameters defined by functions. Ph.D Thesis, University of Missouri, Columbia

Donkin WT, Clark HH (1929) The electric telemeter and valve-spring surge. Trans SAE 24:185–196

Jehle F, Spiller WR (1929) Idiosyncrasies of valve mechanisms and their causes. Trans SAE 24:197–212

Le L, LinY (1994) Stress analysis and optimal cross-section design of noncircular spring wire. Mach Elem Mach Dyn ASME 71

Lin Y, Hodges PH Pisano AP (1993) Optimal design of resonance suppression helical springs. Trans ASME, J Mech Des 115:380–384

Lin Y, Pisano AP (1987) General dynamics equations of helical springs with static solution and experimental verification. J Appl Mech 54:910–917

Lin Y, Pisano AP (1988) The differential geometry of the general helix as applied to mechanical springs. J Appl Mech 55:831–836

Lin Y, Pisano AP (1990) Three-dimensional dynamic simulations of helical compression springs. Trans ASME, J Mech Des 112:529–537

Love EH (1944) Treaties on the mathematical theory of elasticity, 4th edn. Dover, New York

Pearson D (1982) The transfer matrix method for the vibration of compressed helical springs. J Mech Eng Sci 24(4):163–171

SAE (1997) Manual on design and application of helical and spiral. HS-79548-51

Shimoseki M, Hamano T, Imaizumi T (2003) FEM for springs. Springer, New York

Spring Manufacturers Institute (2002) Handbook of spring design.

Tai KK, Lin Y, Wolansky EB (1997) Derivation and experimental verification of design formulae for barrel-shaped helical springs. Springs 36(1)

Wikipedia (2005) Castigliano's method. http://en.wikipedia.org/wiki/Castigliano%27s_method. Accessed 12 Dec 2011

Wittrick WH (1966) On elastic wave propagation in helical springs. Int J Mech Sci 8:25–47

Xu G, Ko P. H, Du R (2011) A study on the precision of mechanical watch movement with tourbillon. Sound Vib 330(26):6287–6544

Zhang YH, Liu HH (1997) Spring handbook (in Chinese), Chinese Mechanical Engineering Press, Ningbo pp 407–418

Chapter 5
Automatic Winding Device

5.1 A Historical Review of the Automatic Winding Devices

The automatic winding device (often called the automatic device in the watch and clock industry) is another fascinating device that is found only in the mechanical watch movement. Its first appearance can be dated back to 1770. The original idea might be attributed to a Swiss watchmaker, Abraham-Louis Perrelet (Wikipedia 2006), but successful applications were found decades later. Perrelet's design used an oscillating weight that moves up and down during the walking and hence, was rather boggy. In 1780, Abraham-Louis Breguet made another design, but was not very reliable. In 1923, John Howard invented the swinging oscillating weight, which became the foundation of the modern automatic watch movements. In 1930, Rolex Company improved this design and used in its Oyster Perpetual. Due in part to this invention, the company raised to become a leader in the watch industry.

The core idea of the automatic winding device is to collect energy through the arm movements of the wearer. As shown in Fig. 5.1, the energy produced by the movement of wearer's arm is captured using an asymmetric oscillating weight (the rotor) and transmitted through a reduction gear train to wind up the mainspring (Reymondin et al. 1999). The external movement, i.e., the movement of the arm, is the energy source. With the arm swinging, the oscillating weight rotates to drive a gear train and finally winds the mainspring in the barrel storing the potential energy. Throughout the years, various designs have been developed to improve its effectiveness, such as using different shapes and materials for the oscillating weight, using ball bearings to reduce the friction, etc. The early designs would capture energy when the rotor is rotating in one direction, usually clockwise. The later designs could work on bidirections.

From a mathematical point of view, as shown in Fig. 5.1 (Xie et al. 2009), this system can be regarded as a double pendulum, where the swinging arm is the

R. Du and L. Xie, *The Mechanics of Mechanical Watches and Clocks*,
History of Mechanism and Machine Science 21, DOI: 10.1007/978-3-642-29308-5_5,

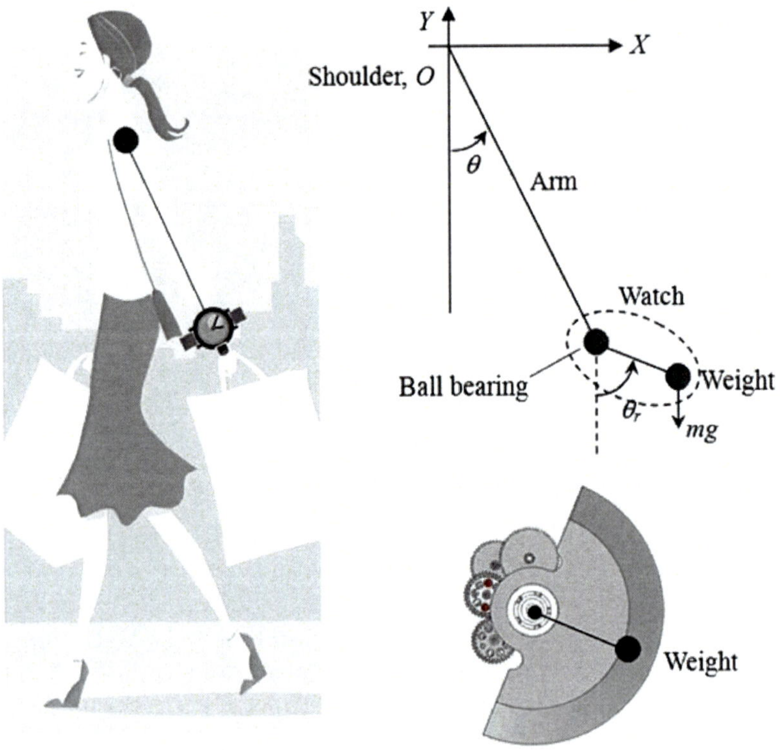

Fig. 5.1 The working of an automatic winding device

upper pendulum pivoted at the shoulder (point O), and the oscillating weight is the lower pendulum pivoted at the oscillating bearing. Note that the movement is in a 2D plane.

According to literature and market survey, there are several different designs currently in production, including the ETA design, the Rolex design and the Seiko design. What distinguishes them from each other is the mechanism named "reverser." It is embedded between the oscillating weight and the reduction gear train so as to enable the automatic device to winding the mainspring no matter which direction the oscillating weight rotates.

It shall be mentioned that although the automatic winding device has been used for many years, few have studied its kinematics. In this chapter, two types of automatic winding devices are studied in details. Section 5.2 studies the ETA automatic winding device and Sect. 5.3 studies the Seiko automatic winding device. Based on the automatic winding device, Sect. 5.4 presents a couple of energy-harvesting devices that can be used to power mobile electronics. Finally, Sect. 5.5 contains conclusions.

5.2 The ETA Automatic Winding Device

5.2.1 The CAD Model

The ETA automatic winding device is perhaps the most commonly used design in the industry today. Figure 5.2 shows its CAD model in front and back view. It mainly comprises an asymmetric oscillating weight (the rotor) mounted on a bearing and a gear train to transmit the motion to the mainspring; both of them are mounted onto a framework (the automatic device lower bridge) to keep their relative positions. When the wearer's arm moves, the oscillating weight swings around its pivot center (it is also the center of the mechanical watch movement) due to its moment of inertia. Note that the oscillating weight contains two parts: the support part must be flexible to absorb the received shocks, whereas the weighted part is responsible to produce large rotation moment. Thus, it is usually made of two different materials: The support part is made of steel, whereas the weighted part is made of tungsten, and the two parts are jointed together by riveting. The rotation movements are transmitted to the gear connected to the ball bearing. The ball bearing on the oscillating weight is important as it reduces the friction. The key is the reverser design as shown in Fig. 5.3. It includes four gears (the reversing wheel, the reversing coupling wheel, the auxiliary reserving wheel and the auxiliary reversing coupling wheel) and four pawls, which transfers the bidirectional movement of the oscillating weight to the unidirectional movement of the reversing pinion. This device also has a stop mechanism, which includes the ratchet wheel and the ratchet wheel driving wheel, to prevent the wounded mainspring from turning back. Finally, when the mainspring is completely wound, a friction-based mechanism in the barrel keeps it from being over-wounded, which has been discussed in Chap. 4.

Figure 5.3 shows the reversing mechanism. It is designed to convert the bidirectional swinging of the ball bearing of the oscillating weight to the unidirectional rotation of the reduction wheel connect to the reversing pinion. As shown in the figure, the reversing mechanism consists of the auxiliary reversing wheel, the auxiliary reversing coupling wheel, the reversing coupling wheel, the reversing wheel and the four pawls. When the oscillating weight turns counterclockwise (see the solid arrows in Fig. 5.3), the auxiliary reversing coupling wheel turns clockwise and drives the auxiliary reversing wheel turning clockwise through the pawls. The auxiliary reversing wheel then drives the reversing wheel and the reversing pinion turning counterclockwise. On the other hand, the reversing coupling wheel also turns; though, its multi-cam profile makes the pawls slide over. As a result, the reversing wheel will not move. In other words, the reversing coupling wheel is effectively decoupled from the reversing wheel at this moment. Similarly, when the oscillating weight turns clockwise (see the *dashed arrows* in Fig. 5.3), the auxiliary reversing coupling wheel turns counterclockwise, but its multi-cam profile makes the pawls slide over. As a result, the auxiliary reversing coupling wheel and the auxiliary reversing wheel are decoupled. On the other hand, the

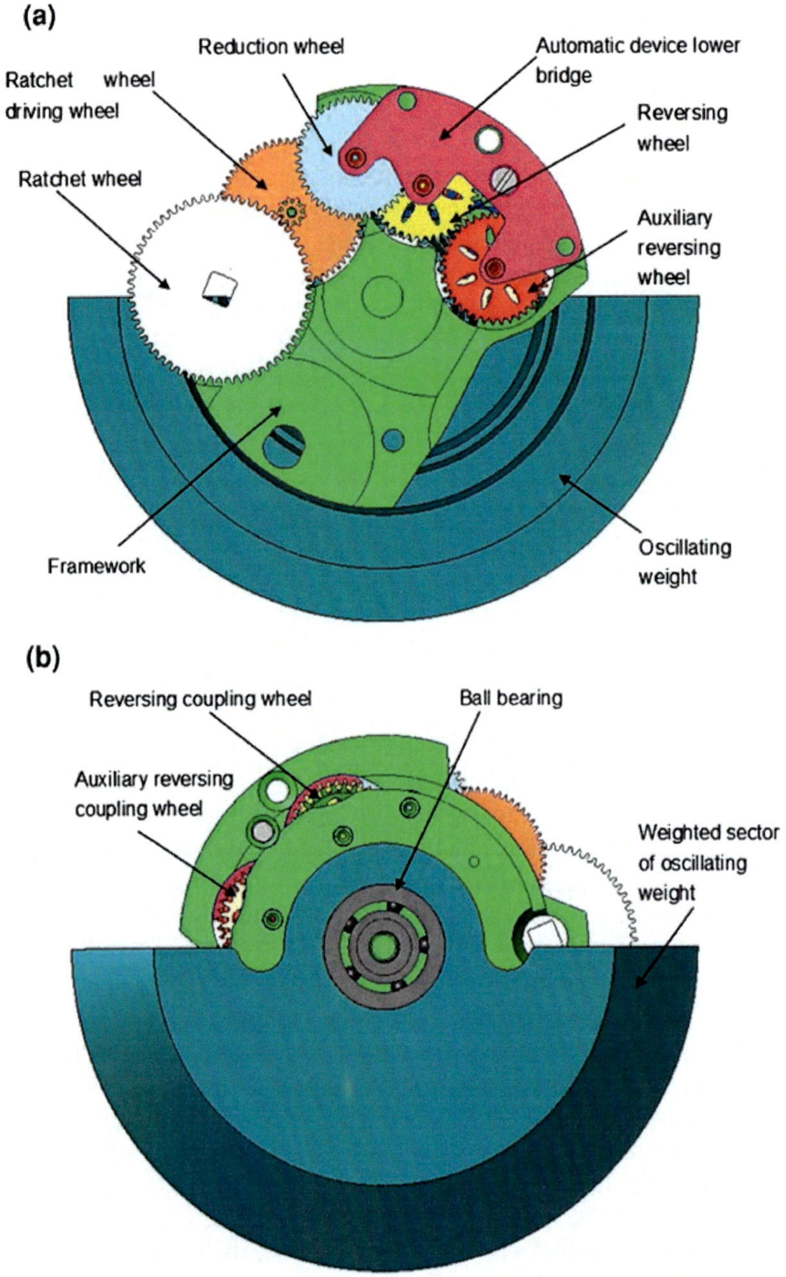

Fig. 5.2 The CAD model of the automatic winding device. **a** The front view **b** The back view

reversing coupling wheel turns counterclockwise and drives the reversing wheel, and the reversing pinion turns counterclockwise through the pawls. Therefore, in both cases, the reversing pinion turns in counterclockwise direction.

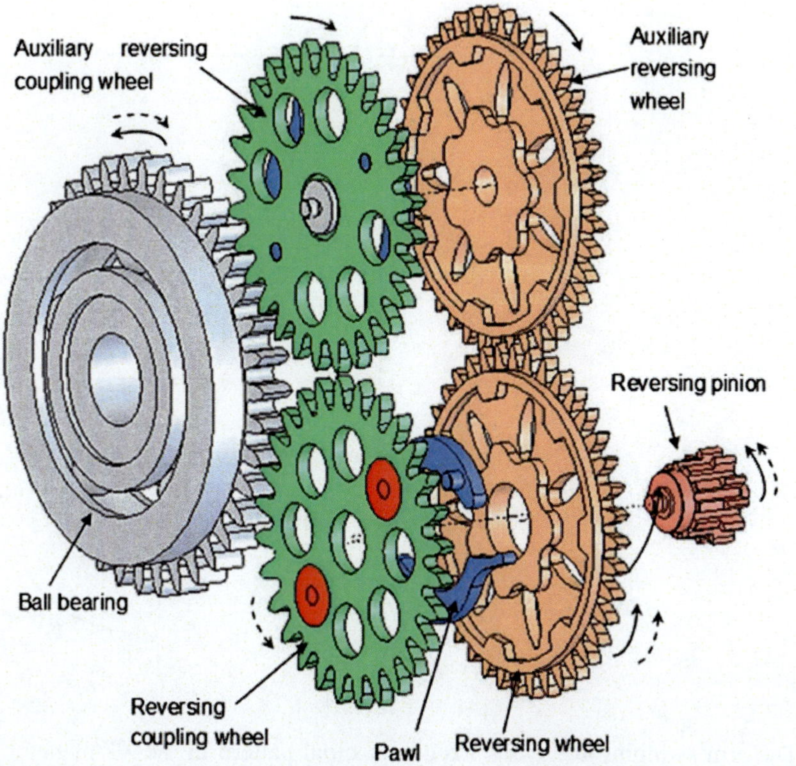

Fig. 5.3 The reversing mechanism of the ETA automatic winding device

In order for the automatic winding device to function well, a reduction gear train is used to increase the torque for winding the mainspring. Starting from the reversing pinion, the reduction gear train includes a reduction wheel, a ratchet driving wheel and a ratchet wheel (refer to Fig. 5.2a). Usually, the reduction ratio of the ratchet and the oscillating weight is between 1:110 and 1:180. The ratchet wheel transmits the movement and the torque to wind the mainspring. When the ratchet wheel turns by a certain angle, the mainspring turns the same angle as well.

A computer animation can be found in the Website of the book http://www.ipe.cuhk.edu.hk/projects10_vl=10English.html.

5.2.2 The Kinematics Model

From a physical point of view, the automatic winding device can be modeled by a double pendulum in the *XY* plane. As shown in Fig. 5.4 (also refer to Fig. 5.1), the swinging arm can be considered as the upper pendulum pivoted at the shoulder, and the rotor (the oscillating weight) can be considered as the lower pendulum pivoted at the ball bearing. For simplicity, following assumptions are made:

Fig. 5.4 Schematic diagram of winding system when arm is swinging

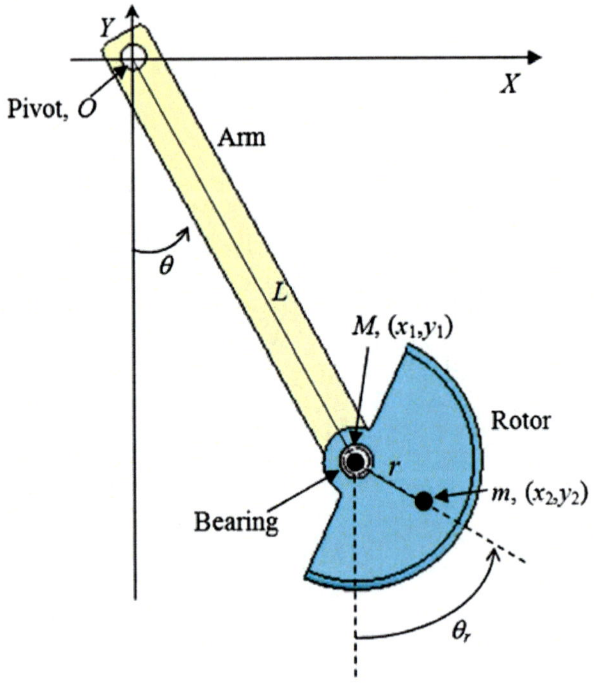

- The arm swinging follows a fixed sinusoidal pattern in the *XY* plane;
- The friction of the bearing is negligible; and
- There is no disturbance during the arm swinging.

Furthermore, for convenience, following symbols are introduced:
M is the mass of the upper pendulum;
m is the mass of the rotor;
L is the length of the upper pendulum;
r is the distance from the center of rotor to the center of mass of the rotor;
(x_1, y_1) is the coordinate of the end of arm, which is also the pivot of rotor;
(x_2, y_2) is the coordinate of the center of mass of the rotor;
θ is the swinging angle of the arm (0 is vertical downward, and positive in counter-clock direction)
θ_r is the swinging angle of rotor (0 is vertical downward, and positive in counter-clock direction).

The upper pendulum rod can be regarded as massless and rigid with a point mass, M, and the rotor can also be considered as a point mass, m. By trigonometry, the following four expressions can be deduced to express the positions (x_1, y_1, x_2 and y_2) in terms of the arm swinging angle, θ, and the rotor rotation angle, θ_r:

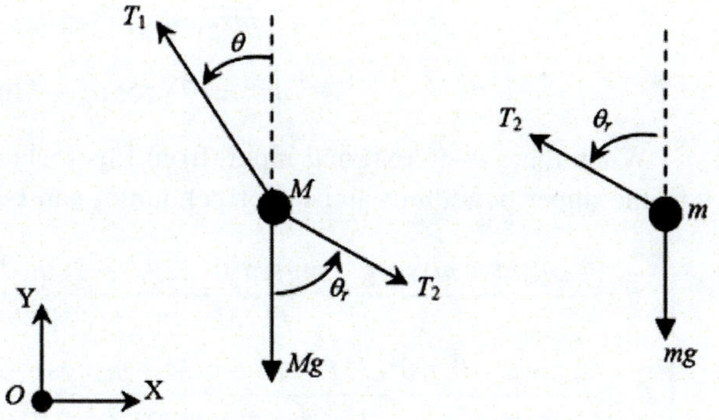

Fig. 5.5 The force diagram of the system

$$\begin{cases} x_1 = L\sin\theta \\ y_1 = -L\cos\theta \\ x_2 = x_1 + r\sin\theta_r \\ y_2 = y_1 - r\cos\theta_r \end{cases} \tag{5.1}$$

Subsequently, we have:

$$\begin{cases} \dot{x}_1 = L\dot{\theta}\cos\theta \\ \dot{y}_1 = L\dot{\theta}\sin\theta \\ \dot{x}_2 = \dot{x}_1 + \dot{\theta}_r r\cos\theta_r \\ \dot{y}_2 = \dot{y}_1 + \dot{\theta}_r r\sin\theta_r \end{cases} \tag{5.2}$$

$$\begin{cases} \ddot{x}_1 = -L\dot{\theta}^2\sin\theta + L\ddot{\theta}\cos\theta \\ \ddot{y}_1 = L\dot{\theta}^2\cos\theta + L\ddot{\theta}\sin\theta \\ \ddot{x}_2 = \ddot{x}_1 - \dot{\theta}_r^2 r\sin\theta_r + \ddot{\theta}_r r\cos\theta_r \\ \ddot{y}_2 = \ddot{y}_1 + \dot{\theta}_r^2 r\cos\theta_r + \ddot{\theta}_r r\sin\theta_r \end{cases} \tag{5.3}$$

The force diagram of the system is shown in Fig. 5.5. The forces acted on the upper pendulum include the tension from the upper rod T_1, the tension from the lower rod T_2 and the gravity force Mg. The forces acted on the lower pendulum include the tension from the lower rod T_2 and the gravity force mg. According to Newton's law, Eqs. (5.4)–(5.7) can be deduced. Note that both the friction of the upper pendulum and the friction of the lower pendulum are very small and thus, have been neglected.

$$M\ddot{x}_1 = -T_1\sin\theta + T_2\sin\theta_r \tag{5.4}$$

$$M\ddot{y}_1 = T_1\cos\theta - T_2\sin\theta_r - Mg \tag{5.5}$$

$$m\ddot{x}_2 = -T_2 \sin\theta_r \tag{5.6}$$

$$m\ddot{y}_2 = T_2 \cos\theta_r - Mg \tag{5.7}$$

When there is no external input, from Eqs (5.1)–(5.7), the angular acceleration of the upper pendulum and lower pendulum can be derived as follows:

$$\begin{cases} \ddot{\theta} = \dfrac{-g(2M+m)\sin\theta - mg\sin(\theta-2\theta_r) - 2\sin(\theta-\theta_r)m(\dot{\theta}_r^2 r + \dot{\theta}^2 L\cos(\theta-\theta_r))}{L(2M+m-m\cos(2\theta-2\theta_r))} \\ \ddot{\theta}_r = \dfrac{2\sin(\theta-\theta_r)(\dot{\theta}^2 L(M+m) + g(M+m)\cos\theta + \dot{\theta}_r^2 rm\cos(\theta-\theta_r))}{r(2M+m-m\cos(2\theta-2\theta_r))} \end{cases} \tag{5.8}$$

It is known that a double pendulum system may undergo chaotic motion and exhibit a motion pattern that is dependent on the initial condition of θ and θ_r. Though, if the initial conditions satisfies, the following inequality holds (Wikipedia 2002),

$$3\cos\theta + \cos\theta_r \geq 2 \tag{5.9}$$

Then, it is energetically impossible for either pendulum to flip, that is, the system must be stable. For the automatic winding device, the initial values of θ and θ_r are all close to zero (since the arm is usually in the rest position) and hence easily satisfies Eq. (5.9); therefore, it will not get into chaotic motion.

Furthermore, the upper pendulum (the swinging arm) acts as the energy input, whereas the lower pendulum, the rotor, is the output. Assume the input θ follows a sinusoidal function, $\theta = a\sin(\omega t)$, where a is the amplitude to the arm swinging; then the output function θ_r can be solved from the following equation:

$$\begin{cases} \theta = a\sin(\omega t) \\ \ddot{\theta}_r = \dfrac{2\sin(\theta-\theta_r)(\dot{\theta}^2 L(M+m) + g(M+m)\cos\theta + \dot{\theta}_r^2 rm\cos(\theta-\theta_r))}{r(2M+m-m\cos(2\theta-2\theta_r))} \end{cases} \tag{5.10}$$

Note that in Eq. (5.10), only θ_r is unknown and hence, can be solved numerically.

As for the gear train as shown in Fig. 5.6, it is assumed that the friction between gears is negligible, the ratio of the torque acted on the first axis (the bearing) and the torque acted on the last axis (the ratchet) can then be expressed as follows:

$$\frac{M_{\text{bearing}}}{M_{\text{ratchet}}} = \frac{F_{\text{bearing}} r_{\text{bearing}}}{F_{\text{ratchet}} r_{\text{ratchet}}} = \frac{Z_1 \cdot Z_3 \cdot Z_5 \cdot Z_7}{Z_2 \cdot Z_4 \cdot Z_6 \cdot Z_8} \tag{5.11}$$

where Z_i is the number of teeth and $i = 1, 2, \ldots, 8$ is the index of Gear i. Based on the data in Wikipedia (2006), one pair of gears alone loses about 10% of the energy. Thus, the overall efficiency of the gears between the oscillating gear and

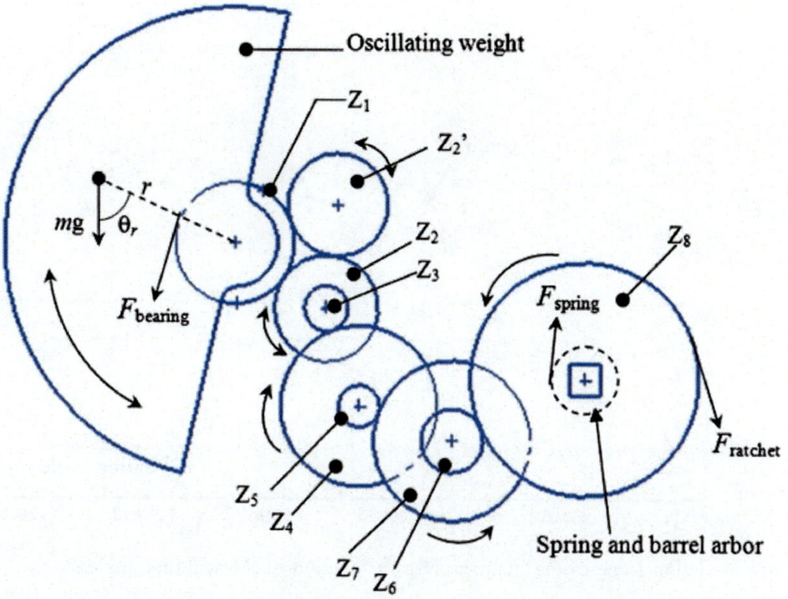

Fig. 5.6 The gear train of the automatic winding device

the ratchet wheel will be $\eta_{\text{total}} = \eta^4 \approx 0.9^4 = 0.656$. Therefore, the actual torque acted on the ratchet wheel can be expressed as follows:

$$M_{\text{ratchet}} = M_{\text{bearing}} \cdot \frac{Z_1 \cdot Z_3 \cdot Z_5 \cdot Z_7}{Z_2 \cdot Z_4 \cdot Z_6 \cdot Z_8} \cdot \eta^4 \tag{5.12}$$

Furthermore, the mainspring is as an energy accumulator. During the winding process, the winding force from the oscillating weight must counter the reaction force of the already wound spring. The reaction force gradually increases as the mainspring is being wounded. Figure 5.7 shows the reaction force of the mainspring, which was collected from experiments using a specially designed instrument called the Variocouple®. In the figure, the vertical axis is the reaction force in mN and the horizontal axis is the winding angle of the mainspring. The two curves are the winding curve (upper) and the unwinding curve (lower). To wind the mainspring, the torque acted on the ratchet wheel must be greater than the reaction torque of the mainspring and hence, the following inequality must be held:

$$M_{\text{spring}} < M_{\text{ratchet}} \tag{5.13}$$

Using Fig. 5.7, the torque of the spring, M_{spring}, can be calculated by multiplying the reaction force, F_r, and the radius of the barrel arbor, r_{arbor}. Furthermore, from Fig. 5.5, the following expression can be obtained:

$$mgr \sin \theta_r = M_{\text{bearing}} = F_{\text{bearing}} \times r_{\text{bearing}} \tag{5.14}$$

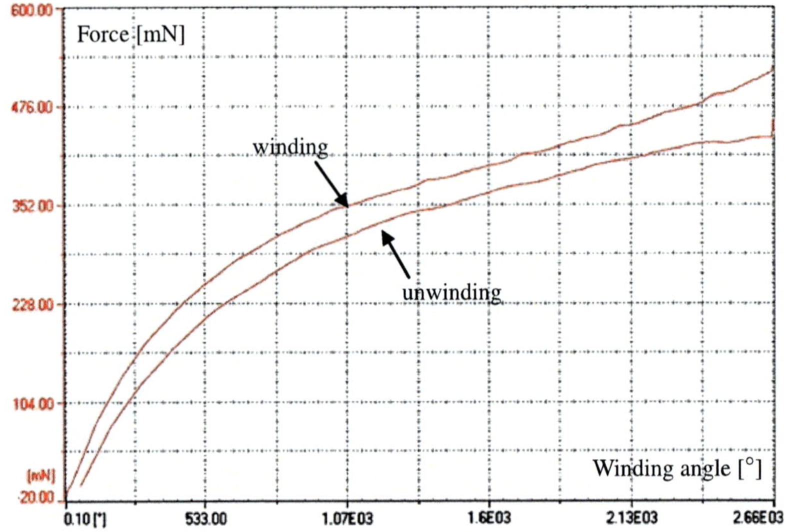

Fig. 5.7 Reaction force of the mainspring as a function of the winding angle

Combining Eqs. (5.12)–(5.14), it follows that:

$$M_{\text{spring}} = F_r \cdot r_{\text{arbor}} < mgr \sin\theta_r \cdot \frac{Z_1 \cdot Z_3 \cdot Z_5 \cdot Z_7}{Z_2 \cdot Z_4 \cdot Z_6 \cdot Z_8} \cdot \eta^4 \tag{5.15}$$

Or

$$\sin\theta_r > \frac{F_r \cdot r_{\text{arbor}}}{mgr \cdot \eta^4} \cdot \frac{Z_2 \cdot Z_4 \cdot Z_6 \cdot Z_8}{Z_1 \cdot Z_3 \cdot Z_5 \cdot Z_7} \tag{5.16}$$

This equation shows the minimum winding angle of the oscillating weight, $\theta_{r,\,\min}$, required to wind the mainspring. From Fig. 5.7, it is seen that $\theta_{r,\,\min}$ should be within $(-\pi/2,\ \pi/2)$; therefore, the minimum angle of the oscillating weight to drive the mainspring can be calculated by:

$$\theta_{r,\,\min} = \arcsin\left(\frac{F_r \cdot r_{\text{arbor}}}{mgr \cdot \eta^4} \cdot \frac{Z_2 \cdot Z_4 \cdot Z_6 \cdot Z_8}{Z_1 \cdot Z_3 \cdot Z_5 \cdot Z_7}\right) \tag{5.17}$$

Finally, the effective angle to wind the mainspring can be obtained by multiplying the ratio of the gear train with the difference between θ_r and θ, $\theta_{r,\,\min}$:

$$\Delta_{\text{eff}}\theta = \max\{(\theta_r - \theta - \theta_{r,\,\min}),\ 0\} \frac{Z_1 \cdot Z_3 \cdot Z_5 \cdot Z_7}{Z_2 \cdot Z_4 \cdot Z_6 \cdot Z_8} \cdot \eta^4 \tag{5.18}$$

5.2.3 Computer Simulation and Experimental Validation

To validate the aforementioned model, computer simulation is conducted. The arm swing (i.e., the movement of the upper pendulum) is assumed to follow a sinusoidal curve, in which the swinging period is 5.19 s and the swinging angle is 27° (0.471 rad). The angular velocity is a constant at 1.21 rad/s. The simulation program is written using MATLAB®. Solving Eq. (5.10) with the initial conditions $\theta_r(0) = 0$, $\dot{\theta}_r(0) = 0$, which satisfies the non-chaotic condition in Eq. (5.9), the angular displacements, angular velocities and angular accelerations of the upper pendulum and the lower pendulum can be obtained, as shown as Fig. 5.8.

According to the measured data using Variocouple® (refer to Fig. 5.7), the mainspring of a mechanical movement (ETA2824) is completely wound at the angle of 14.77π, or 46.4 rad. The simulation result indicates that the time to fully wind the mainspring is 61360 s, or about 17.12 hours. A fully wound watch shall work for approximately 42 hours. It should be pointed out that the watch can work without being fully wound. In fact, after a few swings, the watch shall start to work.

Figure 5.9 shows the winding performance in terms of time under different swing amplitudes. From the figure, it is seen that the higher the amplitude (the higher the arm swings), the shorter time to fully wind the mainspring. Also, when the amplitude is small, the influence of the reaction force is significant, and the winding curve is nonlinear. On the other hand, when the amplitude is big, the influence of the reaction force can be neglected, and the winding curve is approximately linear. Figure 5.10 shows the winding performance in terms of time under different swing speeds. It is seen that the shorter the period (the faster the arm swings), the shorter the time to fully wind the mainspring.

Let the relative angular displacement be $\gamma = \theta_r - \theta$. Then, the total input of the system can be calculated using the following equation:

$$E = \int_0^T J_r \ddot{\gamma}\dot{\gamma}dt \tag{5.19}$$

where J_r is the moment of the rotor and T is total winding time. The total input energy in terms of time is shown in Fig. 5.11. As an example, if the upper pendulum swings as $\theta = 0.471\sin(1.21t)(\text{rad})$, then the theoretic energy input is 1.204 J.

The theoretic energy stored in the main spring, on the other hand, can be calculated by the following expression (Wikipedia 2006):

$$E_{\text{store}} = \frac{e \cdot h \cdot l}{6} \cdot \frac{\sigma_{\max}^2}{E_0} \tag{5.20}$$

where E_{store} is the maximum energy stored in the main spring, σ_{max} the acceptable maximum stress of the mainspring, E_0 the modulus of elasticity, e the thickness of

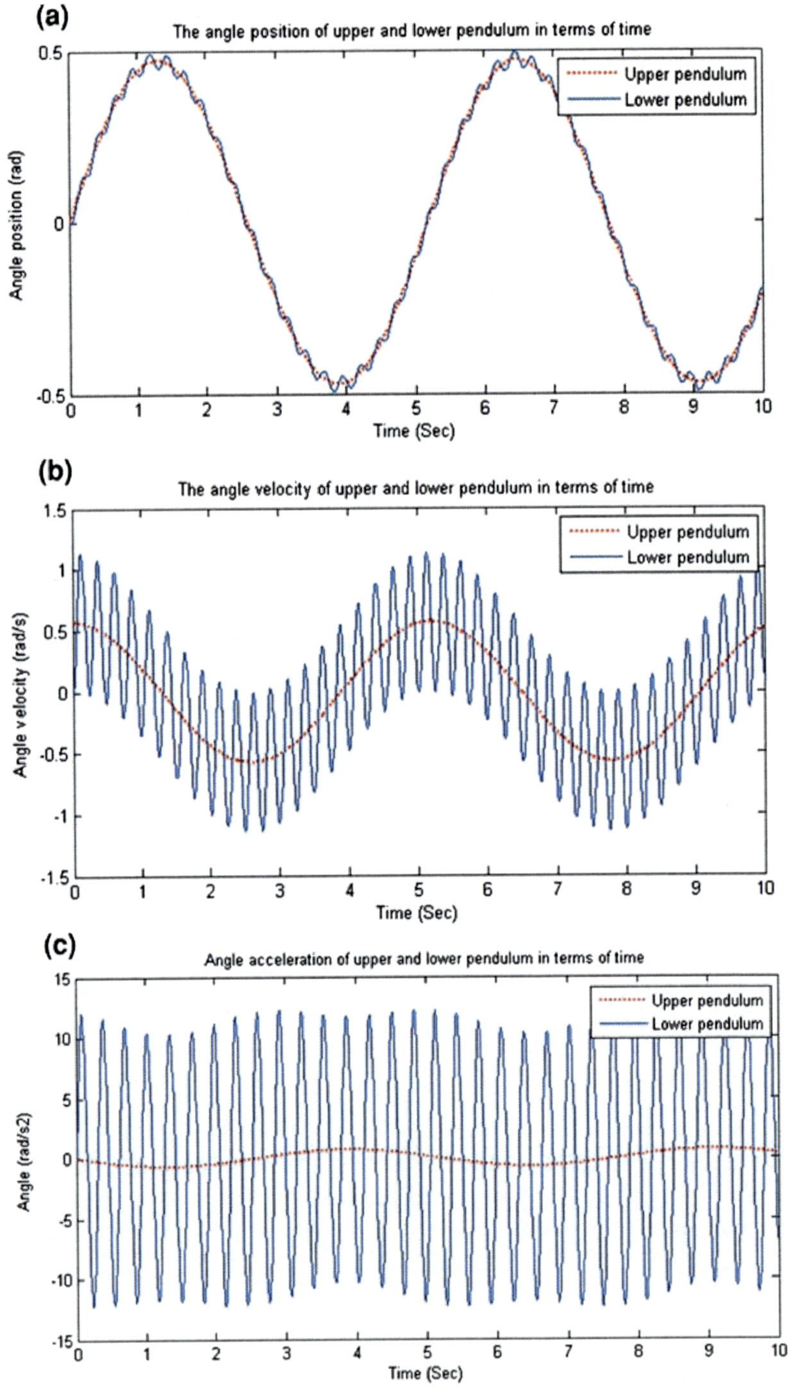

Fig. 5.8 The motion of the upper pendulum and the lower pendulum. **a** The angular displacement **b** The angular velocity **c** The angular acceleration

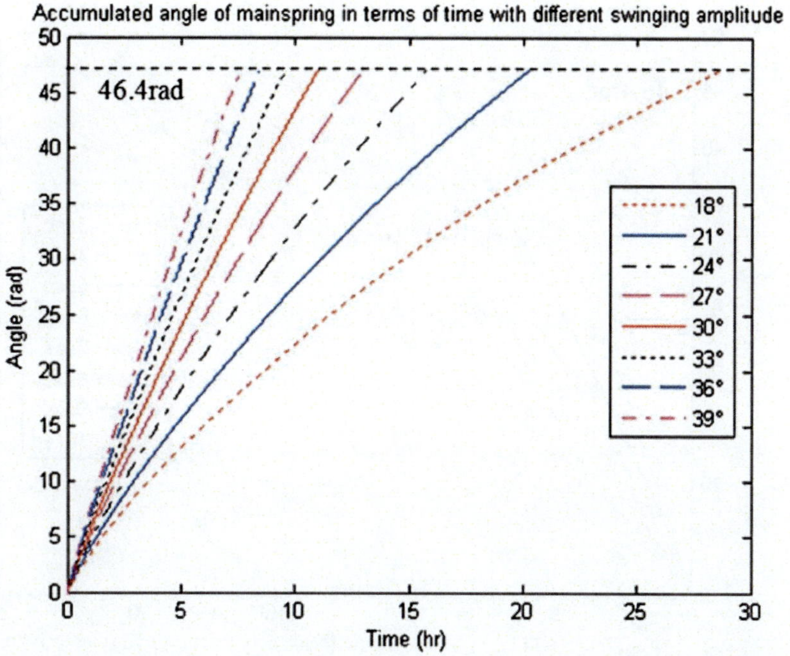

Fig. 5.9 The winding curves under different arm swing amplitudes (the swinging period is set at 5.19 s)

the main spring, h the height of the mainspring and l is the length of the mainspring.

We conducted an experiment using an ETA 2824 mainspring. According to the data given by the manufacturer, $e = 0.134$ mm, $h = 1.23$ mm, $l = 400$ mm, $\sigma_{max} = 3340$ MPa and $E_0 = 2.3 \times 10^5$ MPa, the total energy stored in the mainspring is $E_{\text{stored}} = 557.2\,\text{mJ}$. Therefore, the energy efficiency of the automatic winding system is:

$$\eta_{\text{transfer}} = \frac{E_{\text{store}}}{E_{\text{input}}} \times 100\% = \frac{0.5572}{1.204} \times 100\% \approx 46.3\%$$

The loss of energy is due mainly to the friction of the gears (refer to Eq. (5.12)) and the reversing pawls, as well as to the reaction force of the mainspring.

Different swing speed and swing amplitude would result in different efficiencies. As shown in Fig. 5.12, large swinging speed can produce high efficiency. However, it will saturate at the speed of six times per second. Though, swing amplitude has little effect on the efficiency.

The simulation results are experimentally validated. We built a simple fixture as shown in Fig. 5.13. The watch is fixed on a holder, which is mounted on an aluminum bar driven by a servomotor (the distance between the watch and the motor is 0.5 m). The servomotor can simulate different patterns of arm swing. To observe the winding of the main spring, the ratchet wheel was marked and a

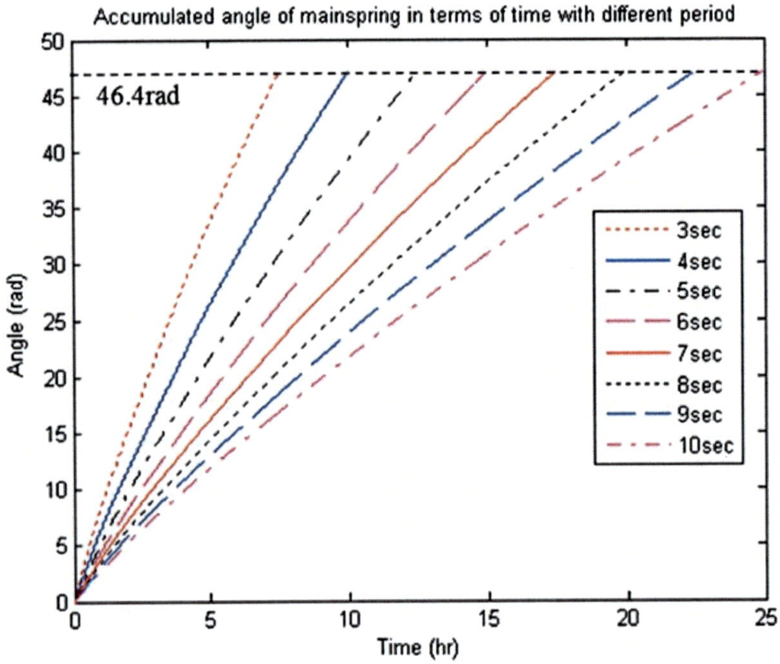

Fig. 5.10 The winding curves under different arm swing periods (the swinging amplitude is set at 27°)

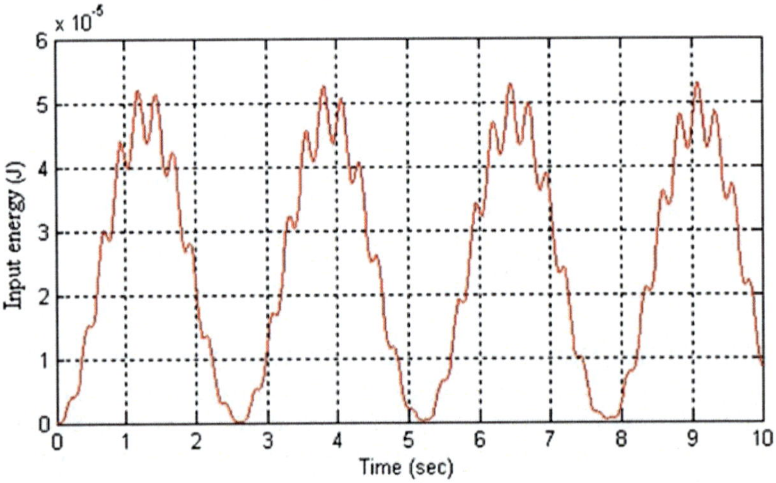

Fig. 5.11 Total input energy in terms of time

camera was fixed on the bar facing the ratchet wheel to monitor its motion. During the experiment, the arm swing is set to 27°. It is found that the total winding time is 19 hours. Comparing to the simulation result of 17.12 hour, the error is less than

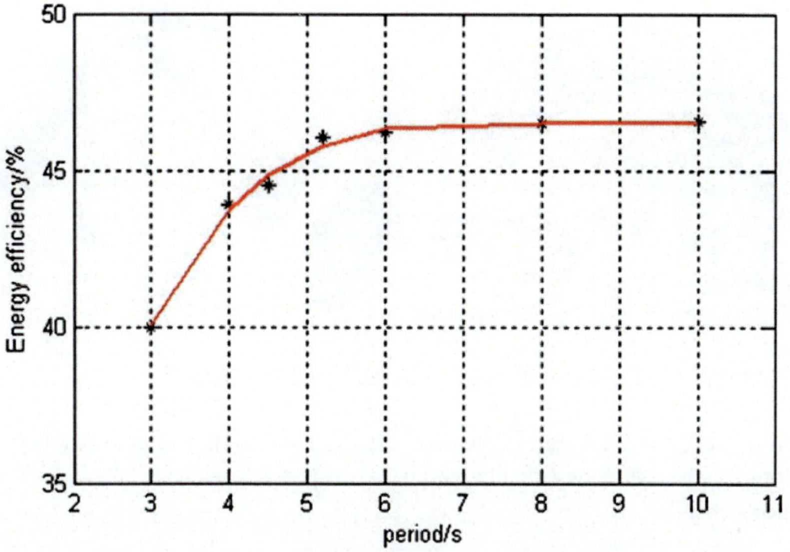

Fig. 5.12 The system efficiency as a function of the swinging period (under a constant swinging amplitude)

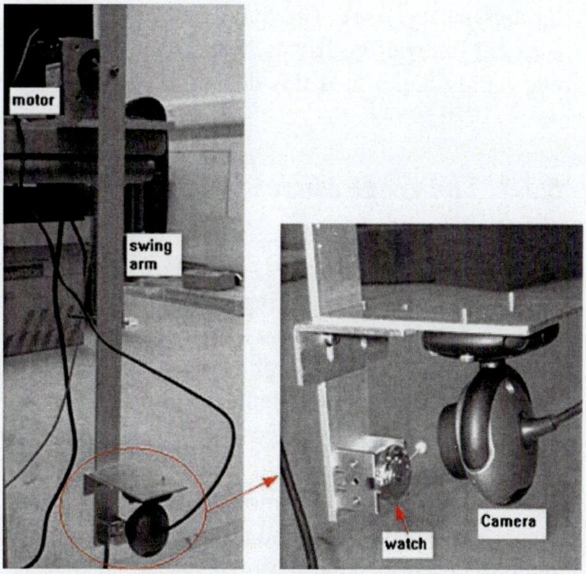

Fig. 5.13 The setup for testing the ETA automatic winding device

10%. The error may be attributed to several reasons. First, many other factors of the watch movement are not considered, such as the energy consumption of the escapement and the gear train. Second, in the simulation model, the friction between the gears is not considered. They both may prolong the winding time.

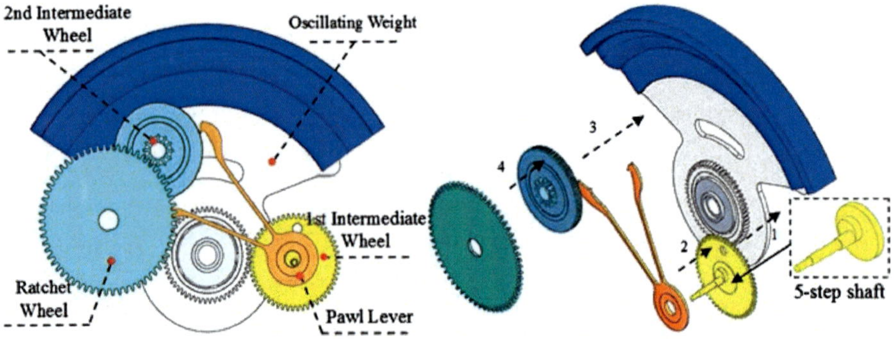

Fig. 5.14 The CAD model of the Seiko automatic winding device

Above all, the ETA automatic winding device is rather effective. According to literatures, the efficiency of nearly all existing energy-harvesting devices hardly exceeds 10%.

5.3 The Seiko Automatic Winding System

The ETA automatic winding device was developed decades ago and refined throughout the years. The Seiko automatic winding device, on the other hand, was a recent innovation. It was inspired by the motion of the locomotive train. Presently, only Seiko uses this design.

5.3.1 The CAD Model

Figure 5.14 shows the CAD model of the Seiko automatic winding device. It is a pawl-lever system consisting of five components: the oscillating weight, the first intermediate wheel, the pawl lever, the second intermediate wheel and the ratchet wheel. The oscillating weight is made of a weighted sector, an oscillating weighted support and a pinion with ball bearing. The first intermediate wheel includes a gear and a five-step shaft. The pawl lever is the reverser, the core of the mechanism. It is mounted on the eccentric shaft of the first intermediate wheel and drives the tooth of the second intermediate wheel when moving. The second intermediate wheel consists of a gear and a concentric pinion. Finally, the ratchet wheel with the pinion is used to wind the mainspring.

Similar to the ETA automatic winding device, the Seiko automatic winding device can work in bidirection. As the arm moves, the oscillating weight drives the second intermediate wheel to rotate in only one direction regardless of the rotating direction of the oscillating weight as shown in Fig. 5.15. When the oscillating weight rotates clockwise, the first intermediate wheel rotates counterclockwise. As a result, as shown in Fig. 5.15a, the pawl lever rotates counterclockwise on the

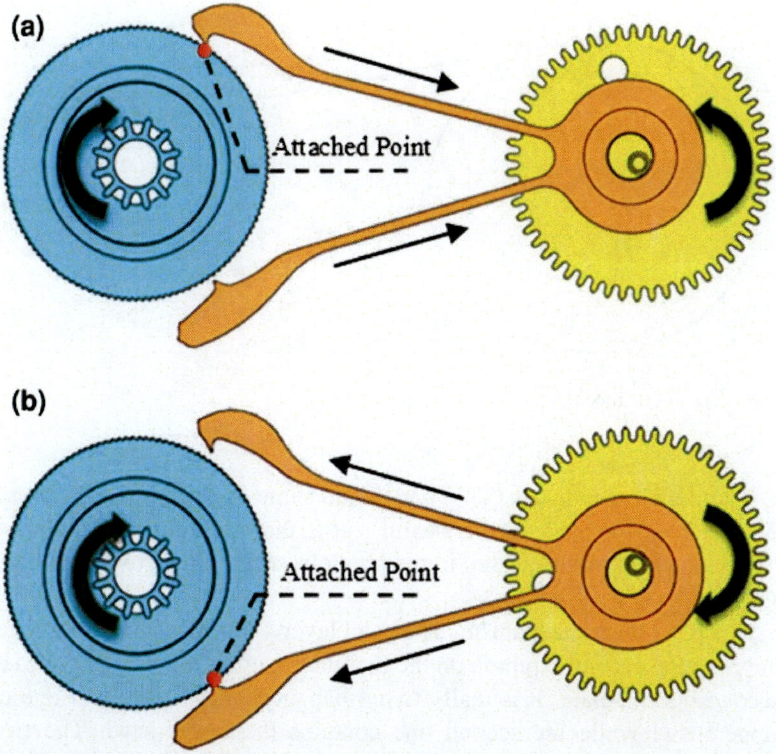

Fig. 5.15 The working Seiko automatic winding device. **a** The oscillating weight rotating clockwise. **b** The oscillating weight rotating counterclockwise

circular path following the eccentric axis of the oscillating weight. Thus, the upper pawl of the pawl lever pulls the tooth of the second intermediate wheel, which drives the second intermediate wheel to rotate clockwise. At this time, the lower pawl of the pawl lever is disengaged. On the other hand, when the oscillating weight rotates counterclockwise, as shown in Fig. 5.15b, the lower pawl of the pawl lever pulls the tooth of the second intermediate wheel while the upper pawl of the pawl lever is disengaged. Thus, the second intermediate wheel rotates clockwise. Consequently, the ratchet wheel only rotates counterclockwise.

A computer animation can be found in the Website of the book http://www.ipe.cuhk.edu.hk/projects10_vl=10English.html or in YouTube at: http://www.youtube.com/watch?v=q5DGLQwCd1M.

5.3.2 The Kinematical Model

The Seiko automatic winding system can be divided into three sub-systems: the double pendulum, the pawl lever 4-bar mechanism and the gear train.

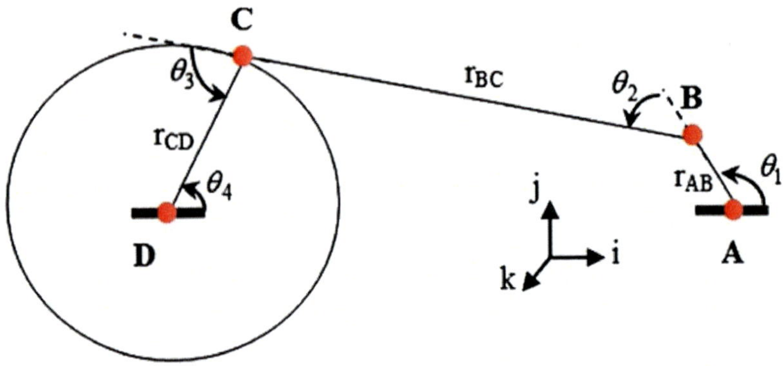

Fig. 5.16 The 4-bar mechanism with the upper pawl

(a) The double pendulum. This part is the same as that of the ETA automatic winding device. It is made of the wearer's arm, the oscillating weight and the first intermediate wheel. As discussed in the previous section, the governing equation is Eq. (5.10).

(b) The pawl lever mechanism. The pawl lever 4-bar mechanism is the reverser. It comprises the eccentric pinion on the first intermediate wheel, the pawl lever and the second intermediate. It is really two 4-bar mechanisms: the first one contains the upper pawl while the second one contains the lower pawl. The two 4-bar mechanisms are similar and hence, we need only to focus on one. Figure 5.16 shows the 4-bar mechanism made of the first intermediate wheel, r_{AB}, the upper pawl, r_{BC}, and the second intermediate wheel, r_{CD}.

To simplify the model, following assumptions are made:

(1) The motion is only in the *XY* plane;
(2) There is no friction in the joints;
(3) The trajectory of Point C is approximately a circle; and
(4) There is no impact force when Point C is in contact with the second intermediate wheel.

For clarity, following notations are adopted:

$\vec{r}_{AB}$—Vector from *A* to *B*, the distance is equal to the difference between the radius of the eccentric pinion and the radius of the axis of the first gear;
$\vec{r}_{BC}$—Vector from *B* to *C*, the distance is equal to the length from the center of the eccentric circle of the pawl lever to the edge of the second intermediate wheel;
$\vec{r}_{CD}$—Vector from *C* to *D*, the distance is equal to the radius of the second intermediate wheel;
θ_1—Angle between $\vec{r}_{AB}$ and the direction of $\vec{i}$;
θ_2—Angle between $\vec{r}_{BC}$ and $\vec{r}_{AB}$;
θ_3—Angle between $\vec{r}_{CD}$ and $\vec{r}_{BC}$;
θ_4—Angle between $\vec{r}_{CD}$ and the direction of $-\vec{i}$;

$\vec{v}_m$, $m = A, B, C$—Velocities of Points A, B and C;
$\vec{a}_m$, $m = A, B, C$—Accelerations of Points A, B and C;
ω_m, $m = A, B, C$—Angular velocities of the bars AB, BC and CD; and
α_m, $m = A, B, C$—Angular accelerations of the bars AB, BC and CD;

It can be shown that the kinematic model of the 4-bar mechanism is as follows:

$$\begin{bmatrix} r_{BC}\sin(\theta_1+\theta_2) & r_{CD}\sin(\theta_1+\theta_2+\theta_3) \\ r_{BC}\cos(\theta_1+\theta_2) & r_{CD}\cos(\theta_1+\theta_2+\theta_3) \end{bmatrix} \begin{bmatrix} \alpha_{BC} \\ \alpha_{CD} \end{bmatrix} =$$
$$\begin{bmatrix} -r_{AB}\cos\theta_1 - \frac{r_{AB}^2}{r_{BC}}\frac{\sin^2(\theta_2+\theta_3)\cos(\theta_1+\theta_2)}{\sin^2\theta_3} - \frac{r_{AB}^2}{r_{CD}}\frac{\sin^2\theta_2\cos(\theta_1+\theta_2+\theta_3)}{\sin^2\theta_3} & -r_{AB}\sin\theta_1 \\ r_{AB}\sin\theta_1 + \frac{r_{AB}^2}{r_{BC}}\frac{\sin^2(\theta_2+\theta_3)\cos(\theta_1+\theta_2)}{\sin^2\theta_3} + \frac{r_{AB}^2}{r_{CD}}\frac{\sin^2\theta_2\sin(\theta_1+\theta_2+\theta_3)}{\sin^2\theta_3} & -r_{AB}\cos\theta_1 \end{bmatrix} \begin{bmatrix} \dot{\theta}_1^2 \\ \ddot{\theta}_1 \end{bmatrix} \tag{5.21}$$

Furthermore, the angular accelerations of the joint angles are:

$$\vec{\ddot{\theta}}_2 = \vec{\alpha}_{BC} - \vec{\ddot{\theta}}_1 \tag{5.22}$$

$$\vec{\ddot{\theta}}_3 = \vec{\alpha}_{CD} \tag{5.23}$$

$$\vec{\ddot{\theta}}_4 = \vec{\alpha}_{CD} \tag{5.24}$$

Given the initial conditions of four joints (including the angular displacement, the angular velocity and the angular acceleration), the movements of the four joints can then be solved.

(c) The gear train. Similar to the ETA automatic winding device, the effect of the gear train can be modeled by the transmission ratio. There are two pairs of gears in the Seiko's automatic winding device, so the total efficiency of the gear train is $\eta_{total} = \eta^2$.

It should be pointed that although the pawl lever 4-bar mechanism can describe the interactions among the first intermediate wheel, the pawl lever and the second intermediate wheel, there is also a "jump" when the first intermediate wheel moves, which shall be taken into consideration. As shown in Fig. 5.16, Point C is the touch point of the upper pawl lever against the second intermediate wheel. During the operation, the upper pawl lever, BC, moves along the teeth of the second intermediate wheel. Accordingly, the effective radius of the second intermediate wheel varies. Therefore, Eq. (5.21) only describes part of the motion. To accommodate this variation, a modification is necessary. Figure 5.17 shows the block diagram for computing the motion of the second intermediate wheel. This computation procedure handles the rotation of the first intermediate wheel in both directions. Also, it should be mentioned that when the pawl lever reaches its furthest point on the second intermediate wheel, it tends to move backward slightly, while the latter does not follow but continues to move at a constant velocity until the pawl lever moves forward again. This is considered as negligible and is not included in the model.

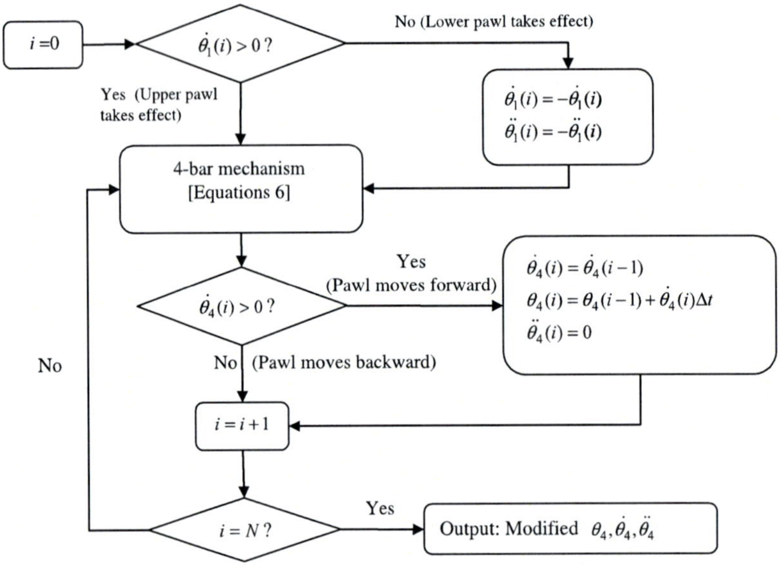

Fig. 5.17 The procedure for computing the motion of the second intermediate wheel

Using the procedure in Fig. 5.17, the angular displacement, angular velocity and angular acceleration of the ratchet wheel can be obtained by simply multiplying θ_4, $\dot{\theta}_4$, $\ddot{\theta}_4$ with the gear ratio and the resistant force from the wounded mainspring.

5.3.3 Computer Simulation

The aforementioned model is implemented using MATLAB®. Figure 5.18 shows the flowchart of the computer program. The differential equation system is solved using the fourth-order Runge-Kutta method. The integration interval is 0.01 s, and the integration time is 10 s covering a few arm swings.

Two case studies are presented herein.

(a) *Case 1—normal walking*. The arm is swinging in a sinusoidal curve, $\theta = \frac{\pi}{4}\sin\left(\frac{2\pi}{3}t\right)$, starting from the rest position. The swing amplitude is 45° and the frequency is $2\pi/3$ Hz. Figure 5.19 shows the behavior of the oscillating weight and the ratchet wheel (i.e., the mainspring). From the figure, it is seen that the period of the oscillating weight is about 0.142 s. In other words, the oscillating weight swings approximately 21 times within one arm swing. Additionally, the swing amplitudes are different. Furthermore, the angular velocities of the ratchet wheel are continuously increasing. This implies that the ratchet wheel rotates in the same direction. The angular displacement of the ratchet wheel represents the accumulated rotation of the mainspring. From the simulation result, it is seen that the

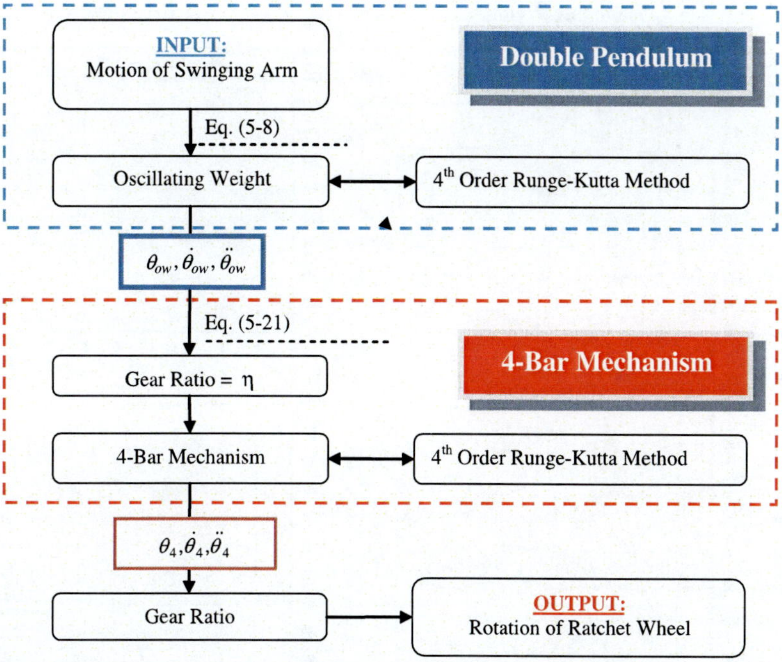

Fig. 5.18 The flowchart of the simulation

ratchet wheel rotates a total of 140 degree, which is resulted from only four arm swings.

(b) *Case 2*. Raising the arm to read the time. Assume that a person takes 2 s to raise his or her hand with a constant speed and then stops when his or her arm reaching 90° from his or her thigh, the angular displacement/velocity of the arm can be described as follows:

$$\theta = \begin{cases} 0, & t = 0 \\ \frac{\pi}{4}t, & 2 > t > 0 \\ \frac{\pi}{4}, & t > 2 \end{cases}$$

$$\dot{\theta} = \begin{cases} 0, & t = 0 \\ \frac{\pi}{4}, & 2 > t > 0 \\ 0, & t > 2 \end{cases}$$

That is, the starting position is the rest position. In the first 2 s, the arm swings upward with a constant speed of $\pi/4$ rad/s. After 2 s, the velocity of the arm swing becomes 0, while the arm remains at 45°. Figure 5.20 shows the behavior of the oscillating weight and the ratchet wheel. In the first 2 s, the movement is a

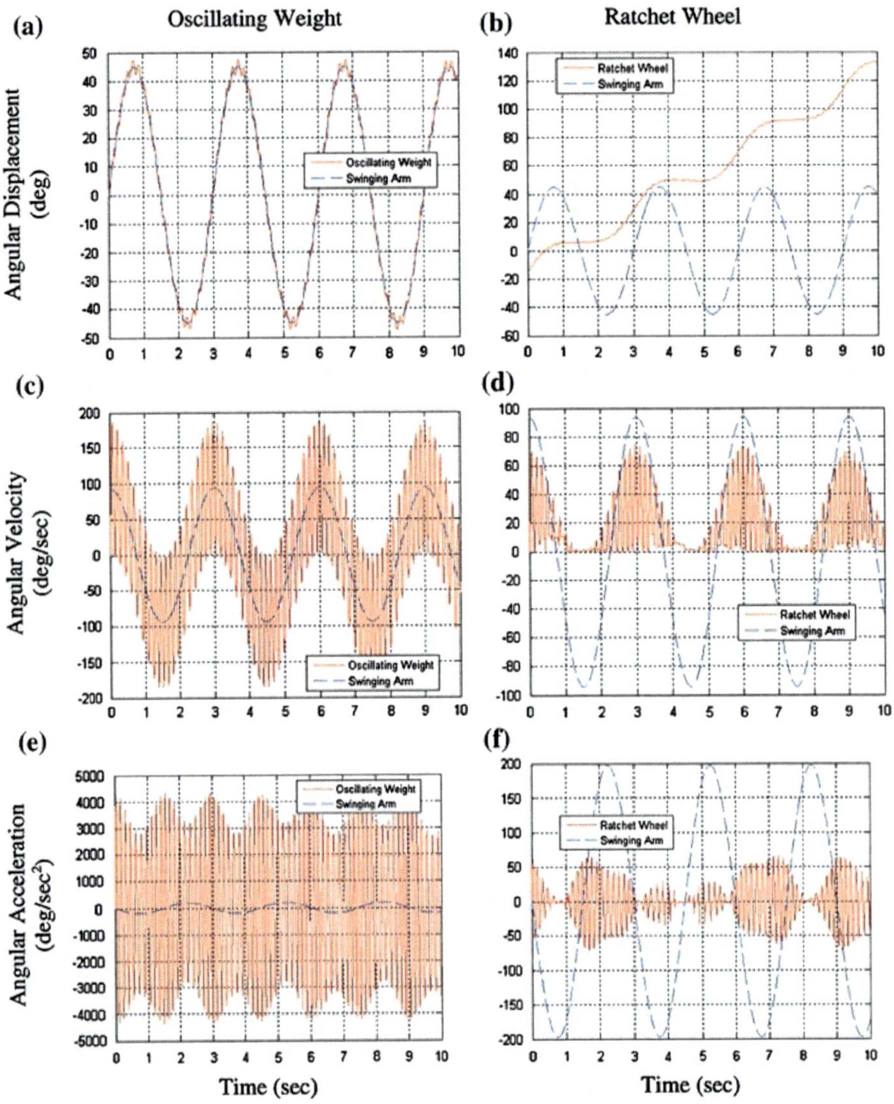

Fig. 5.19 Simulation result in Case 1

transient movement; after that, the movement is a periodic motion. Note that the movement of the oscillating weight will eventually die out because of the friction and the reaction force from the mainspring. Though, it is not shown in the figure because of the short simulation time.

The effects of the swing amplitude and frequency are also studied. Figure 5.21 shows the simulation results, in which the input is the amplitude of the arm swing and the output is the amplitude of the ratchet wheel. As shown in Fig. 5.21a, the input/output ratio is almost the same, which implies that the motion of the ratchet wheel is proportional to the motion of arm. The input/output ratio is the indication

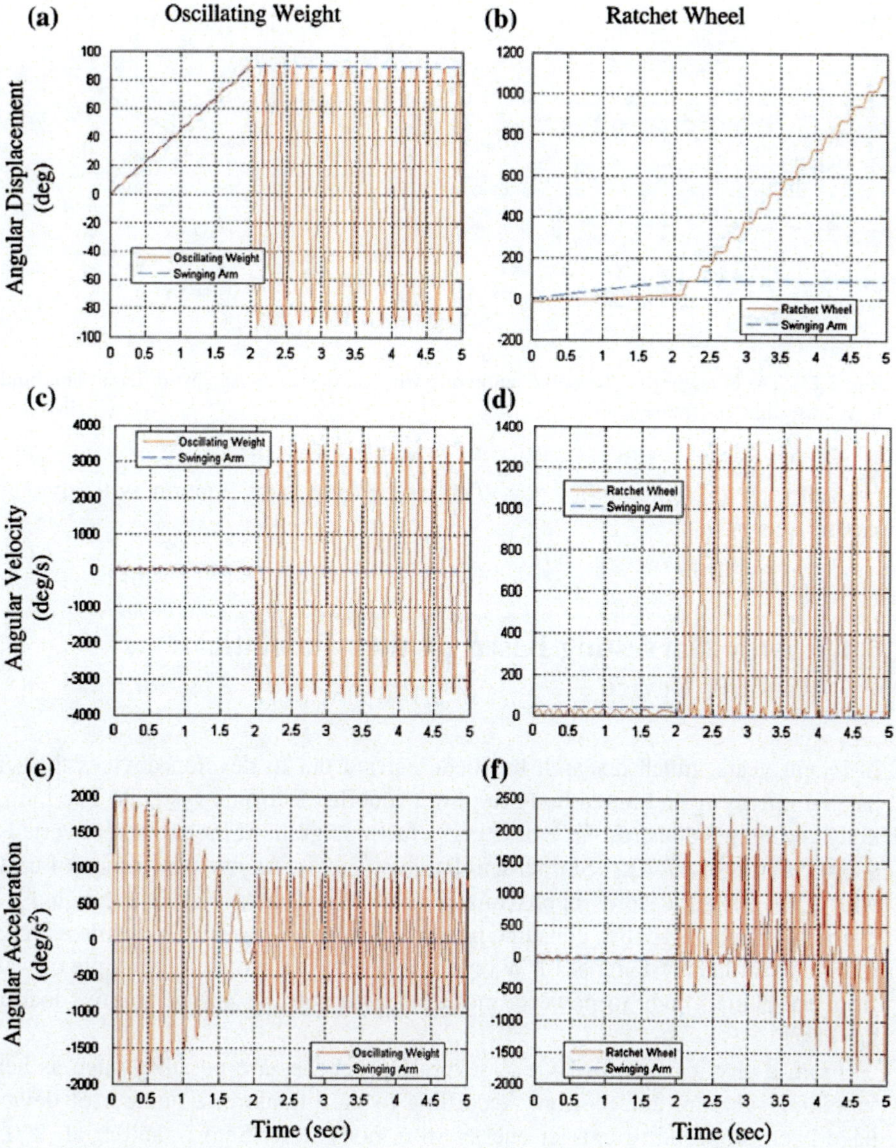

Fig. 5.20 Simulation result in Case 2

of the efficiency. The figure shows that the efficiency of the Seiko automatic winding device is approximately 42%, which is slightly less than that of the ETA automatic winding device of 46%.

In addition, as shown in Fig. 5.21b, when the arm swing frequency increases, the ratio increases linearly with a slope of 0.261. It implies that the high arm swing

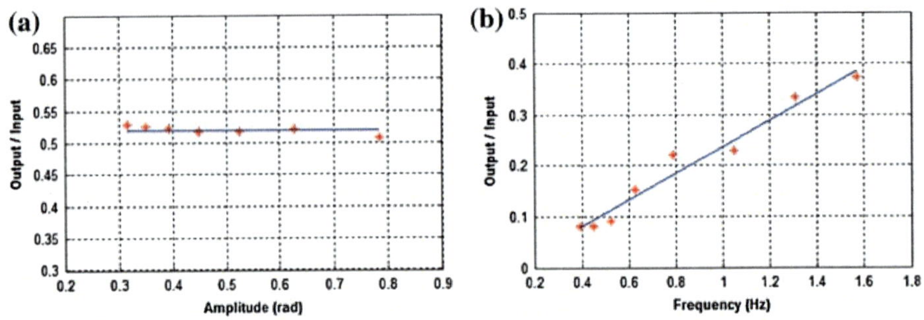

Fig. 5.21 The efficiency of the Seiko automatic winding device. **a** Against different amplitudes **b** Against different frequencies

frequency gives better efficiency. This is because slow motion will give less kinetic energy.

5.4 Energy Harvesting Based on the Automatic Winding Device

In recent years, much research has been carried out to develop devices that can harvest energy from human body to power mobile electronic systems. The human body is a tremendous resource of energy. Just one gram of fat can be converted to 9,000 calories or 37.7 kJ (Starner and Paradiso 2004). An average person of 68 kg with 15% body fat stores approximately 384 MJ of energy. It is estimated that everyday human activities consume power at a rate of 81–1,630 W, while a typical mobile electronic system uses 1 W or less. Therefore, scavenging a couple of watts from the human body to power a mobile device will not put an onerous load on the user.

From a physical point of view, several approaches are possible, such as heat, vibration, pressure and motion. According to the literature, a number of devices have been proposed to extract energy from body heat (Sato N and et al. 2005), walking (Turri et al. 2004), movement-induced vibration (Beeby et al. 2006; Roundy and Wright 2003), shoe pressure (Shenck and Paradiso 2001), breathing (Kornbluh et al. 2002) and typing (Crisan 1999). The most recognized ones are two articles published in *Science*, both of which utilize electromagnetic mechanism to extract kinetic energy from human body motion. The first one (Rome et al. 2005) used a suspended-load backpack as the inertia mass to convert mechanical energy from the vertical movement of the carried loads to electricity. The device can produce 7.4 W in the walking at the speed of 5.6 km/h. However, its load of 38 kg is too much for normal person, and the device can only collect the vibration in vertical direction. The other one (Donelan et al. 2008) also used the

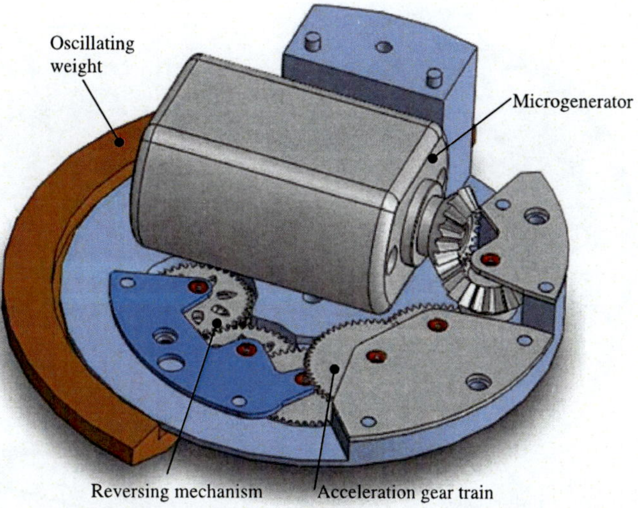

Fig. 5.22 Schematics of our energy harvester (Xie et al. 2009)

electromagnetic mechanism to generate electricity during human walking. This knee brace-like device is supposed to assist muscles in performing negative work, analogous to regenerative braking in hybrid cars, where energy normally dissipated during braking drives a generator instead. It mounts at the knee and selectively engages power generation at the end of the leg swing, thus assisting deceleration of the joint. Test subjects walking with such a device on each leg produced an average of 5 watts of electricity. The weight of such setup is about 1.5 kg, which is obviously inconvenient. Furthermore, it is not clear if human bodies welcome such a foreign device.

The idea of using human body energy to power a quartz watch was also reported (Kanesaka et al. 1999), and there are successful commercial products available in the market. We proposed a more efficient device (Xie et al. 2009) that can harvest kinetic energy to produce electrical power for portable electronic devices, such as cellular phone, GPS receiver and night vision goggle. Figure 5.22 shows the conceptual design of the device. It consists of an oscillating weight, a reverser that changes the bidirectional movements to the unidirectional rotation, an acceleration gear train to speed up the rotation, a micro generator to produce electricity, a control circuit and a battery to store the electricity (not shown in the figure). The major challenges of building this device are to minimize the friction in various parts and to design a micro generator that can effectively handle small inputs.

Our second design, as shown in Fig. 5.23, made some improvement. This harvesting device mainly consists of a rotor, a stator, a torsion spring and a control circuit (not shown in the figure). The eccentric rotor, made of permanent magnet, serves as the oscillating weight to harvest the arm swing. Because of the rotation of

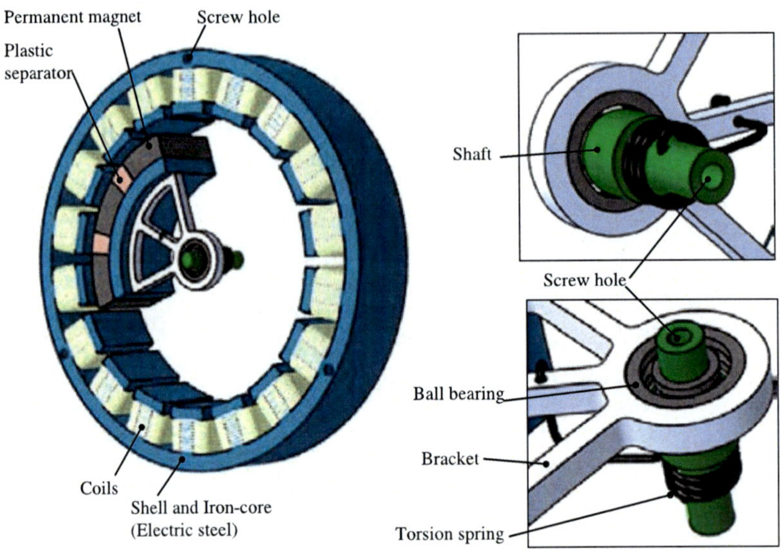

Fig. 5.23 Schematics of the our second energy harvester

the rotor, the permanent magnets produce a changing magnetic field. The coils, wound on the iron core, serves as the stator, thus produces electricity. It is known that the arm swing is not always on a vertical plane, even in normal walking. Thus, the device adds a torsion spring, one end of which is fixed on the bracket of the rotor, the other fixed on the shaft, so the device can work even if the arm swing is on the horizontal plane. Based on the model presented in Sect. 5.2, the device can generate approximately 600 mW in normal walking. It is expected such a device would find many applications in the near future.

5.5 Concluding Remarks

Automatic winding device is another unique device developed for mechanical watch movement. The idea can be dated back to two hundred years ago. There are several different designs in the market today. In this chapter, two designs, the ETA design and the Seiko design, are investigated in detail. From a physical point of view, the automatic winding device can be seen as a double pendulum: the arm is the upper pendulum, whereas the oscillating weight is the lower pendulum. Different designs use different reversers to convert the swing of the oscillating weight to the unidirectional rotation of the mainspring. The ETA design used a pawl level design while the Seiko design used a 4-bar mechanism. Both designs are very effective and their efficiencies are about 45%. This is much higher than that of the energy-harvesting devices reported in the research articles, whose efficiencies are usually only 10%.

Inspired by the automatic winding devices, we designed a new device to power mobile electronic devices, such as a mobile phones, GPS and night vision goggle. The new device may find some applications in the near future.

References

Beeby S, Tudor M, White N (2006) Energy harvesting vibration sources for microsystems applications. Meas Sci Technol, 17:17

Crisan A (1999) Typing power. United States Patent 5,911,529, June 1999

Donelan JM et al (2008) Biomechanical energy harvesting: generating electricity during walking with minimal user effort. Science 319:807–810

Kanesaka T et al (1999) Development of a thermal energy watch. Micromechatronics 43(3):29–36

Kornbluh R et al (2002) Electroelastomers: applications of dielectric elastomer transducers for actuation, generation, and smart structures, smart structures and materials. In: McGowan R (ed) Industrial and commercial applications of smart structures technologies, vol 4698 pp 254–270

Reymondin CA, Monnier G, Jeanneret D, Pelaratti D (1999) The theory of horology. The Technical College of Vallee de Joux, Switzerland

Rome L, Flynn L, Goldman EM, Yoo TD (2005) Generating electricity while walking with loads. Science, 09:172

Roundy S, Wright P, Rabaey J (2003) A study of low level vibrations as a power source for wireless sensor nodes. Comput Commun 26:1131–1144

Sato N et al. (2005) Novel MEMS power generator with integrated thermoelectric and vibrational devices. In: Proceedings of the 13th international conference on solid-state sensors. Actuators and Microsystems, Korea. p 29

Shenck N, Paradiso J (2001) Energy scavenging with shoe-mounted piezoelectric. IEEE Micro 21(3):30–42

Starner T, Paradiso J (2004) Low-power electronics design, chapter 45. CRC Press, New York

Turri S, Miller D, Ben AH, Multon B (2004) Design of an electro-mechanical portable system using natural human body movements for electricity generation. Sens Actuators A: Phys 116(3):461–471

Wikipedia (2002) Double pendulum. http://en.wikipedia.org/wiki/Double_pendulum. Accessed 12 Dec 2011

Wikipedia (2006) Automatic watch http://en.wikipedia.org/wiki/Automatic_watch. Accessed 12 Dec 2011

Xie LH, Menet CG, Ching H, Du RX (2009) Automatic winding device of a mechanical watch movement and its application in energy harvesting. Trans ASME, J Mech Des, 131(3)

Chapter 6
Gear and Power Transmission

6.1 A Historical Review of the Gear Train in Mechanical Movements

The gear is a symbol of engineering. Dating back to 100 B.C., the Greeks already had a good understanding of the gear (Wikipedia 2002). The gears are used to transmit motion and power. There are many kinds of gears: spur gears, helical gears, bevel gears, hypoid gears, crown gears, worm gears, non-circular gears, gear racks, sun and planet gears, harmonic gears, etc. The teeth of a gear may follow different profiles, such as involute, cycloidal, spline, etc. We assume that the reader has basic knowledge of gears and understands the terms, such as module, pitch circle, etc. The reader may also refresh his/her memory by reading reference books, such as (Litvin and Fuentes 2004).

For mechanical watch movements, gears play an essential role. A typical mechanical watch movement has approximately 30 gears carrying out different functions such as winding, setting, etc. The most important ones are those in the main gear train which transfers the kinetic energy from the mainspring (stored in the barrel) to the escapement, as shown in Fig. 6.1. The main gear train has a couple of distinct features. First, it is constantly moving and must move precisely. Second, it is small (typically the wheels are about 1 mm in thickness and the pinions are about 1 mm in diameter) and has a large transmission ratio (e.g., 110/16). As a result, even a small error may jeopardize timekeeping accuracy. In this chapter, we focus on the main gear train.

The gears in the main gear train of the mechanical watch movement are spur gears. Their tooth profile is not the usual involute but cycloid (Litvin and Fuentes 2004). It is not clear who started to use cycloid gears. Compare to the involute gear, the cycloidal gear has two advantages. First, it can accommodate large gear ratio without tooth interference. Second, it can have a small number of teeth. Under the same conditions, the involute gear would become radically undercut,

R. Du and L. Xie, *The Mechanics of Mechanical Watches and Clocks*,
History of Mechanism and Machine Science 21, DOI: 10.1007/978-3-642-29308-5_6,

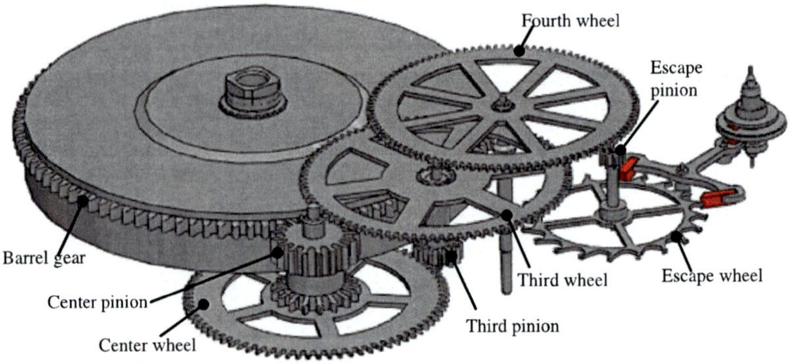

Fig. 6.1 The main gear train in a mechanical watch movement

making it easy to break and difficult to manufacture. Still, owing to their small sizes, cycloid gears are sensitive to manufacturing and/or misalignment errors. Even a small error may cause significant deterioration in performance. Therefore, detailed analysis is helpful.

The gear problem is a complex problem. Before detailed analysis can be carried out, the solid model of the gear pair should be established first. According to the literatures, much effort has been made to construct the gear model. In the late 1980s, the 2D model of an involute spur gear was constructed to study the static contact problem (Varangan and Ganesan 1994). Later, the 3D wire frame model of a spiral bevel gear was developed (Fong and Tsay 1991), and the 3D solid model of a worm gear was also constructed (Bair and Tsay 1998).

Based on the geometric model, the analysis can then be carried out. Usually, the analysis includes the tooth bending and the tooth contact. The tooth bending is relatively simple, as it does not involve motion. The tooth contact, on the other hand, is much more difficult. From the early analytical methods to numerical methods, such as the finite element method (FEM), boundary element method (BEM), finite difference method (FDM), etc., considerable progress has been made. Among these methods, FEM is the most commonly used. For example, in (Bibel et al. 1994), a 2D FEM model of an involute spur gear is developed. In (Greening et al. 1980), FEM is used to show that the load capacity of a double enveloping hourglass worm gear is higher than that of the single enveloping worm gear. In (Simon 1993), a 3D FEA model of a worm-gear is constructed to deal with the contact stress. However, existing tooth contact analysis methods have limitations: Either the contact loads are much simplified or the contact pattern and contact zone are predetermined. Furthermore, misalignment errors are not considered.

In reality, gears, including the spur gear, the involute bevel gear and the worm gear, are all sensitive to misalignment errors, which are inevitable in practice. This is particularly true for gears in mechanical watch movements (Wild 2003). Because of misalignment errors, the contact zone will be shifted and edge contact will occur. To achieve a better contact pattern and decrease the sensitivity of the

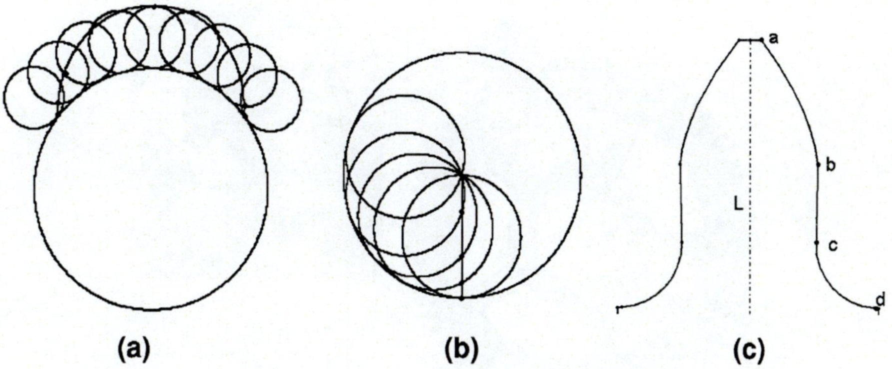

Fig. 6.2 Illustration of the formation of a cycloidal gear profile. **a** Epicycloid. **b** Hypocycloid. **c** Gear profile

gear pair to the misalignment errors, it is desirable to modify the gear profile. For the involute spur gears, Litvin and his team proposed to modify the cut-in depth to control the contact zone (Litvin et al. 1995). They also proposed a similar method for bevel gear and worm gear (Litvin and Lu 1995; Feng and Litvin 1998; Soel and Litvin 1996). It should be pointed out though that the exiting methods are case dependent. Furthermore, they do not get optimal gear profiles. In practice, a careful examination reveals that some industrial companies, such as ETA, use modified gear profiles.

In this chapter, we will study the gear profile and gear profile modifications. It should give the reader sufficient information to design and make gears for mechanical watch movements. The chapter consists of six sections. Section 6.2 presents the geometric design of the cycloidal gear. Section 6.3 shows its FEA model. Section 6.4 shows the FEA simulation, including the bending and contact stresses. Section 6.5 discusses gear profile modification. Finally, Sect. 6.6 contains concluding remarks.

6.2 The Geometrical Model

The cycloidal gear is so-called because its tooth profile is derived from cycloid curves. Cycloid gear profile includes two parts: epicycloid and hypocycloid: the addendum is the epicycloids and the dedendum is the hypocycloid. As shown in Fig. 6.2a, the epicycloid curve is generated by a point at the edge of a circle rolling along another circle outside, whereas the hypocycloid curve, as shown in Fig. 6.2b, is the track of a point at the edge of a circle when rolling along the other circle inside. For mechanical watch movements, the hypocycloidal dedendum is usually a radial straight line (i.e., the radius of the circle is infinite). Figure 6.2c shows a complete tooth profile. The addendum, *ab*, is an epicycloid curve, whereas the dedendum, *bc*, is a hypocycloid curve. They conjugate to form a tooth. To reduce stress concentration, the tooth is filleted generating another curve, *cd*, at the

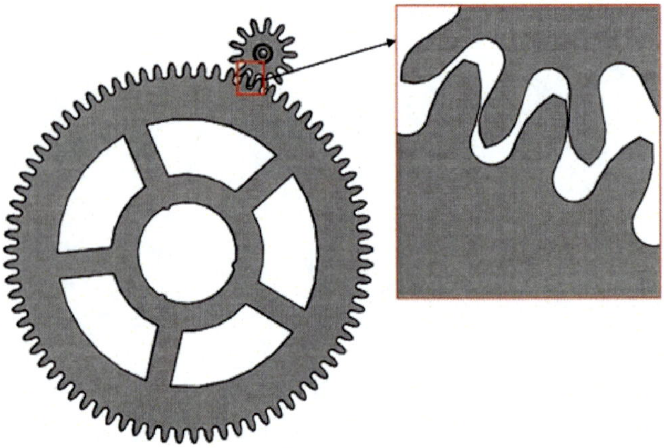

Fig. 6.3 Illustration of the center wheel and the third pinion

bottom. The tooth profile is completed by mirroring about the radial line, L, forming the gear tooth profile.

Now, let us consider a pair of gears: the center wheel and the third pinion, as shown in Fig. 6.3. First, the gear profile can be generated using the British Standard 978 (Soel and Litvin 1996). Let the radius of the rolling circle be r, the center of rotation be r_1 and the distance from rolling center be a, the coordinates on addendum can be generated form epicyloidal Eqs. (6.1) and (6.2). Connecting to the radical straight dedendum and the fillet in the bottom, the gear profile is then formed. Accordingly, the CAD model can be developed as shown in Fig. 6.4. Note that based on the British Standard 978, if the number of teeth is 11 or more, a part of the addendum will be cut out as shown in Fig. 6.4b. In addition, the addendum curve joins the dedendum curve below the base circle. Furthermore, the fillet is a large arc.

$$x = (r_1 + r)\sin\varphi - a\sin\left[\varphi\left(1 + \frac{r_1}{r}\right)\right] \tag{6.1}$$

$$y = (r_1 + r)\cos\varphi - a\cos\left[\varphi\left(1 + \frac{r_1}{r}\right)\right] \tag{6.2}$$

Based on the one-tooth model, the whole gear pair model can then be developed by circular repeat.

6.3 The FEA Model and Simulation Results

Based on the geometrical model, the FEA model can be formed. We use ANSYS® to build two FEA models: one for bending analysis and the other for contact analysis.

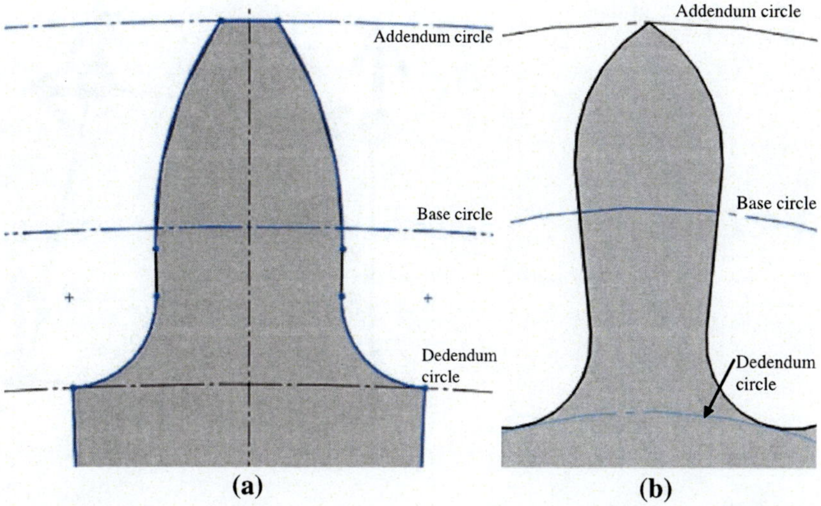

Fig. 6.4 The CAD model a tooth. **a** Profile for gear with 11 teeth or more, **b** Profile for gear with 10 teeth or less

Table 6.1 Material properties of the gear pair

Material	Module of elasticity (N/m^2)	Poisson ratio
Brass	1.103×10^{11}	0.33
Steel	1.896×10^{11}	0.28

6.3.1 The FEA Model for Bending Analysis

Bending analysis is aimed at evaluating the deformation of the gear under load. During the meshing, the gear rotates and its teeth that are involved in the meshing are subject to bending. One may consider that the meshing tooth acts like cantilever resisting the force from the meshing tooth. If the bending is too big, the tooth may fail, usually at the roof of the tooth (Wei 2011). Therefore, FEA is conducted to evaluate the bending.

Following the procedure of FEA, first, material property data are added. It is noted that wheels and pinions are made of brass and steel, respectively. Table 6.1 shows the corresponding material properties.

Next, the CAD model is discretized. This can be done using the automatic mesh generation function of ANSYS®. In particularly, we use PLANE42 element, whose shape function is shown in Eqs. (6.3) and (6.4). This element is defined by four nodes with 2 DOF on each node as shown in Fig. 6.5. The automatic mesh generation is done by using the function SMARTSIZE, which can automatically distribute the mesh to reduce the computation load. Afterwards, refinement can be done on critical areas for more accurate result.

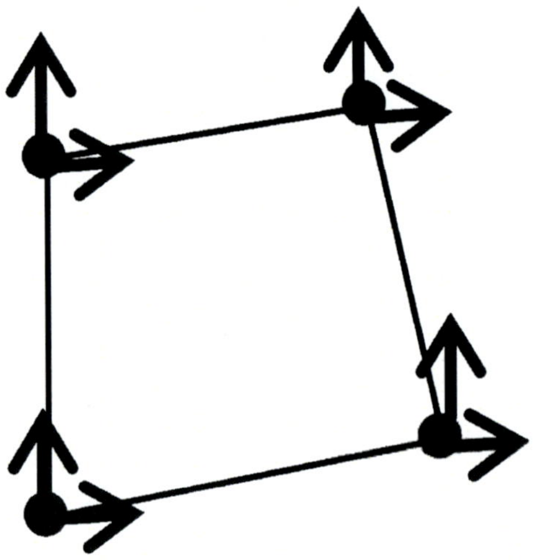

Fig. 6.5 Illustration of the PLANE42 element

$$u = \frac{1}{4}[u_1(1-s)(1-t) + u_2(1+s)(1-t) + u_3(1+s)(1+t) + u_4(1-s)(1+t)] \quad (6.3)$$

$$v = \frac{1}{4}[v_1(1-s)(1-t) + v_2(1+s)(1-t) + v_3(1+s)(1+t) + v_4(1-s)(1+t)] \quad (6.4)$$

Figure 6.6 shows the FEA model of the center wheel (the FEA model of the third pinion can be found in a similar manner). Figure 6.6a is the single-tooth model; note that the mesh is refined at the roof of the tooth near the root that is expected to have large stresses. The boundary conditions of the single-tooth model are shown in the figure as well. First, the DOFs on the sides and the base of the tooth body are removed. Second, a force is applied on one side of tooth tip. This simulates the worst static bending condition (Bibel 2011). Figure 6.6b shows the whole gear model, the worst bending condition is similar to single-tooth model. The center of the wheel is assumed to be hollow to reduce computation load. The internal circle is fixed for initial static condition, while the forces are applied on the tips of three teeth to simulate the worst bending condition including the neighbor influence. Loading magnitude is equal to the torque divided by the radius of the tooth tip, in which the torque is the dividing torque from the previous transmission.

Figure 6.7a and b shows the simulation results (tensile stress) using the single-tooth model and the whole gear model, respectively. The two models exhibit a same pattern. The red zone in the loading side indicates the tension, whereas the blue zone in the opposite side shows the compression. From the figures, it is also seen that the largest stress occurs at the thinnest rim of the gear in the loading side. This will be the most vulnerable place of the gear.

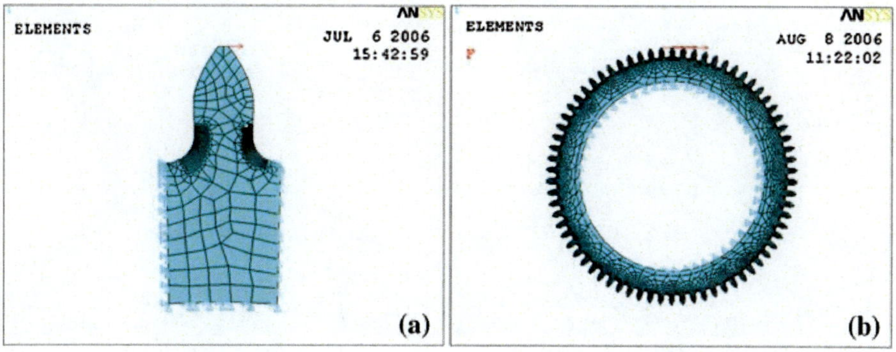

Fig. 6.6 The bending analysis FEA model. **a** Single-tooth model. **b** Whole gear model

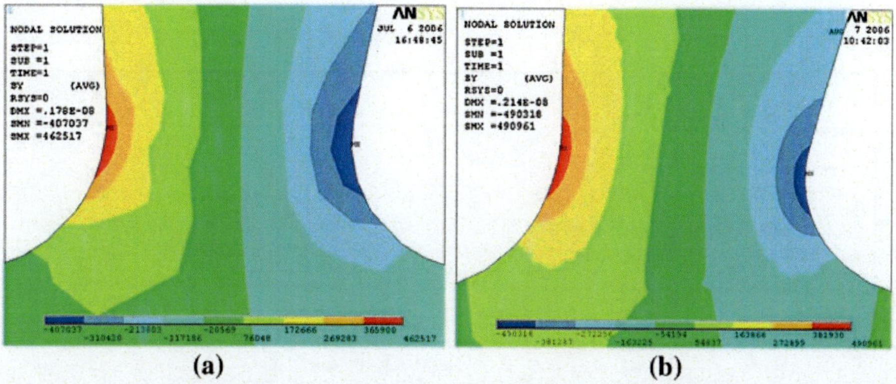

Fig. 6.7 The simulation results (tensile stress) in bending analysis. **a** Using the single-tooth model. **b** Using the whole gear model

As the tensile stress quantifies the crack failure, the von Mises stress quantifies the effective stress. Figure 6.8a and b shows the von Mises stress distribution using the single-tooth model and the whole gear model, respectively. Again, the two models show the same pattern.

Comparing Figs. 6.7 and 6.8, it is seen that the distributions of the tensile stress and the von Mises stress are rather similar, though the amplitudes are different. Figure 6.9 plots the tensile stress in blue and the von Mises stress in purple along the roof of the tooth, where distance 0 is the loading point. It is seen that the two curves are practically overlapping from distance 0 to 0.595 and are symmetry from distance 0.595 to 1.184. This shows that the tensile stress is equivalent to the effective stress. Hence, analyzing the tensile stress is usually enough.

Now, let us consider two pairs of gears in the main gear train: the center wheel—third pinion gear pair and the third wheel—fourth pinion gear pair (refer to Fig. 6.3). These two gear pairs play an important role in timekeeping. Their design parameters (the tooth numbers and the module) are listed in Table 6.2. The

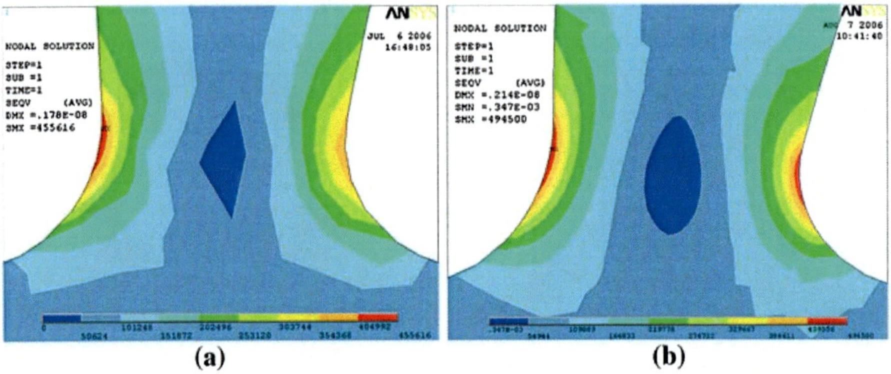

Fig. 6.8 The simulation results (von Mises stress) in bending analysis. **a** The single-tooth model. **b** The whole gear model

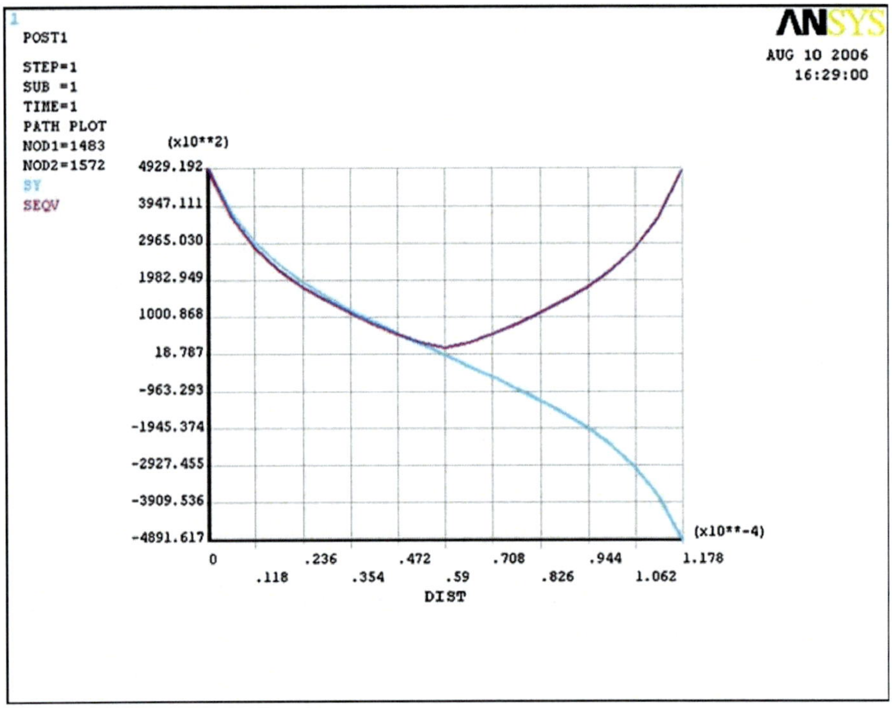

Fig. 6.9 The stress distribution along the roof of the tooth

maximum stresses are shown in Tables 6.3 and 6.4. From the tables, it is seen that the variations between the single-tooth model and the whole gear model are <8%. This indicates that analyzing the single-tooth model is usually enough, which requires much less computational load.

Table 6.2 The design parameters of the gear pairs

Driving gear—driven gear	Tooth number	Module
Center wheel—third pinion	77–13	0.085
Third wheel—fourth pinion	90–9	0.075

Table 6.3 The maximum tensile stresses

	Single-tooth model	Whole model	Variation (%)
Center wheel	0.218	0.219	0.46
Third pinion	4.556	4.899	7.53
Third wheel	0.032	0.033	3.13
Forth pinion	1.623	1.630	0.43

Table 6.4 The maximum von Mises stresses

	Single-tooth model	Whole model	Variation (%)
Center wheel	0.225	0.227	0.89
Third pinion	4.625	4.945	6.92
Third wheel	0.033	0.032	3.03
Forth pinion	1.630	1.628	0.12

From Tables 6.3 and 6.4, it is also seen that the pinions are subject to much larger stresses than that of the wheels. For example, the maximum stress of the forth pinion is 50 times larger than that of the third wheel. Although pinions are made of steel with high yield strength, it still has the risk of failure. Because the bending takes place in every meshing cycle, fatigue must be considered. In other words, choosing high-quality material is very important.

6.3.2 The FEA Model for Contact Analysis

The contact analysis is aimed at evaluating the localized stress on the teeth. As suggested in (Bibel 2011), the gear contact model is a combination of the two gear bending models meshing at the pitch circle, where the relative motion of the gear pair is pure rolling and the friction can be neglected. In the FEA, the two contacting teeth could go past one another without touching. This is because when bending, the teeth act as elastic springs moving away from each other. To deal with this problem, the stiffness relationship between the two contacting areas is modeled by a spring placed in between. In other words, a contact element is inserted between the two contacting areas. In addition, it is essential to prevent the two contacting areas from passing through each other. The method of enforcing contact compatibility is the penalty method (Bibel 2011). It controls the surface penetrations by penalizing the combined normal contact stiffness. Note that the

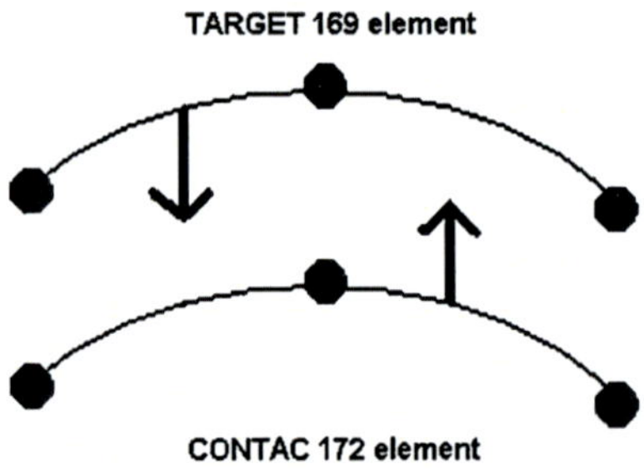

Fig. 6.10 TARGET169 and CONTACT172 element in ANSYS

normal contact stiffness should be carefully selected. If the penalty is too small, the surface penetration may occur; though if the penalty is too large, the combined normal contact stiffness may produce severe numerical problems. In the FEA model, the contact stiffness is the penalty parameter. In almost all cases, the contact stiffness can be automatically determined by ANSYS®. We use the contact stiffness defined below:

$$k = \frac{f_s \times A \times k}{D} \tag{6.5}$$

where A is the area of the contact element, k the bulk modulus of the contacted element, f_s the penalty factor and D is the distance between the two contact elements. This contact relationship is called the flexible-to-flexible contact, where both contacting body are deformable materials. The contact pairs are meshed with surface-to-surface contact elements and do not have restriction on the shape of the surface. Hence, it suits the complicated tooth profile of the gears well. In ANSYS®, under this contact relationship, the driving wheel is modeled with element CONTAC172 and the driven pinion is modeled with element TARGET169 as shown in Fig. 6.10. Note that CONTAC172 is 2D element defined by three nodes with the shape function below (TARGET169 element is similar).

$$W = C_1 + C_2x + C_3x^2 \tag{6.6}$$

In a gear meshing cycle, the contact may occur in various areas of the gear tooth. For simplicity, contact analysis is concentrated at the pitch circle with pure rolling and no friction. This is reasonable under the assumptions of prefect meshing. Figure 6.11 shows the FEA model. Figure 6.11a is the single-tooth contact model of the center wheel driving the third pinion. It is known that contact conditions are sensitive and hence, mesh refinement is taken near the pitch circle and the bottom of the gear. Note that the boundary conditions are applied to both teeth. As the tooth of the wheel is much smaller than the body of the wheel, it is considered as a rigid beam to reduce the computational load. The rigid beam can rotate around the rolling center. Also, the DOF at the bottom of the wheel tooth is removed. The driving torque is simulated by a line loading. The loading is assumed to be uniform and equal to the average force times the length of the tooth. The boundary conditions of the pinion are defined in the similar manner.

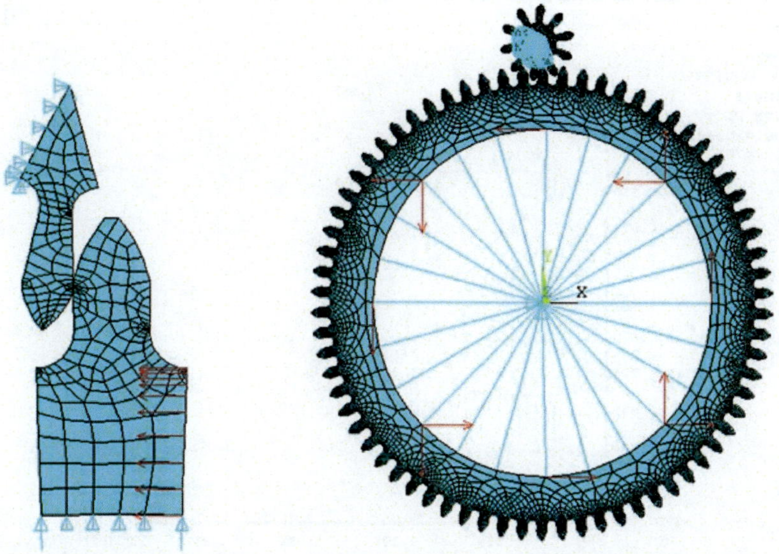

(a) (b)

Fig. 6.11 The contact analysis FEA models. **a** Single-tooth model. **b** Whole gear model

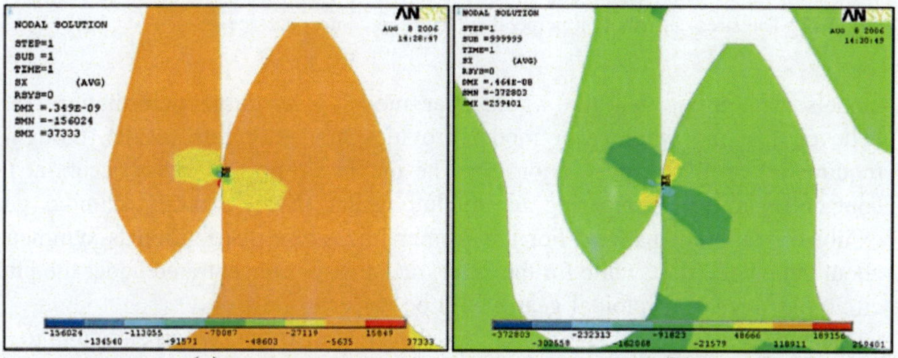

(a) (b)

Fig. 6.12 The simulation results (tensile stress) in contact analysis. **a** Using the single-tooth model. **b** Using the whole gear model

Figure 6.11b shows the whole gear model. The pinion is constrained by removing the DOFs in the inner circle. The body of the wheel is assumed to be hollow and each tooth is considered as a rigid beam as that in the single-tooth model. The center of the wheel is constrained allowing only the rotation. The torque is applied to the body of the wheel in terms of x and y components, which simulates the rolling of the wheel in the counterclockwise direction as that of the single-tooth model.

Figure 6.12a shows the simulation result using the single-tooth model, and Fig. 6.12b shows the simulation result using the whole gear model. From the figure, following observations can be made: (a) The resulting patterns from the two

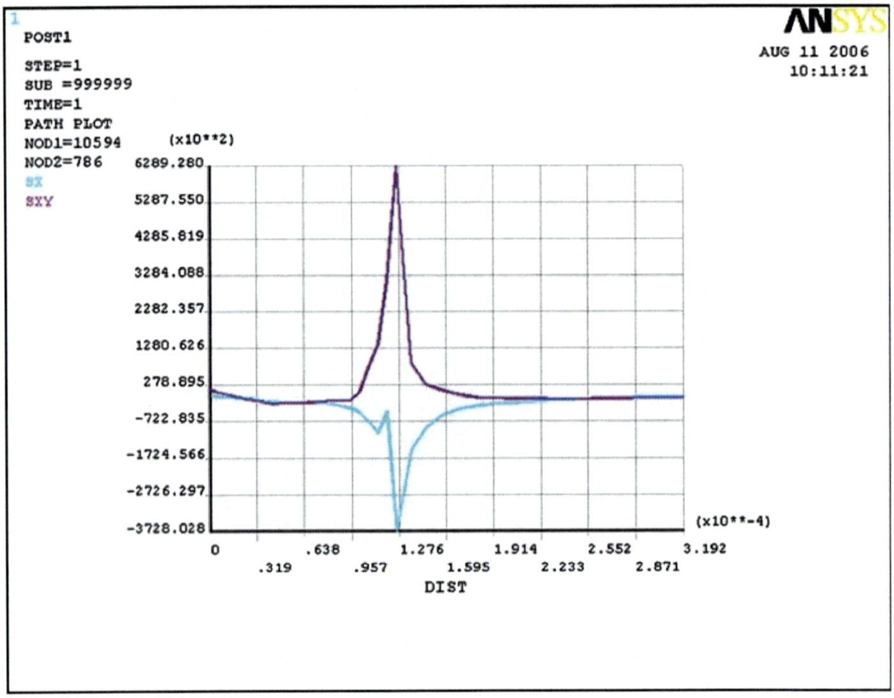

Fig. 6.13 The stress distribution perpendicular to the contact surface

models are similar, but the whole gear model gives large deformation. It is believed that the whole gear model provides more accurate results, though it requires much more computation. (b) The maximum normal stress occurs at the contact point. (c) Comparing to involute gears (Bibel 2011), cycloidal gear exhibits a different pattern. For the former, the stress distribution is symmetric about the pitch circle, while for the latter, the stress is concentrated under the pitch circle. As a result, cycloidal gear would be easier to fail.

To have a better view, Fig. 6.13 shows the stress distribution perpendicular to the contact surface using the whole gear model. The contact point is at position 1.256 with the pinion on the left and the wheel on the right. The curve in *purple* is for the wheel, whereas the curve in light *blue* is for the pinion. From the figure, it is seen that the normal stress is concentrated at the contact surfaces of the teeth. Furthermore, Fig. 6.14 shows the stress distributions of the wheel and the pinion. Form the figure, it is seen that the stress diminishes within micrometers from the contact.

Table 6.5 shows the maximum normal stress of the main gear train of the aforementioned mechanical movement. It is seen that the difference between the two models is <1.5% and hence, is insignificant. Table 6.6 shows the maximum shear stress. This time, the difference between the two models is significant, especially for the center wheel and the third pinion pair. This is due large to the fact that the gear ratio plays a role. With smaller number of teeth in the center wheel than third wheel, the angle of loading line in the single-tooth model of the

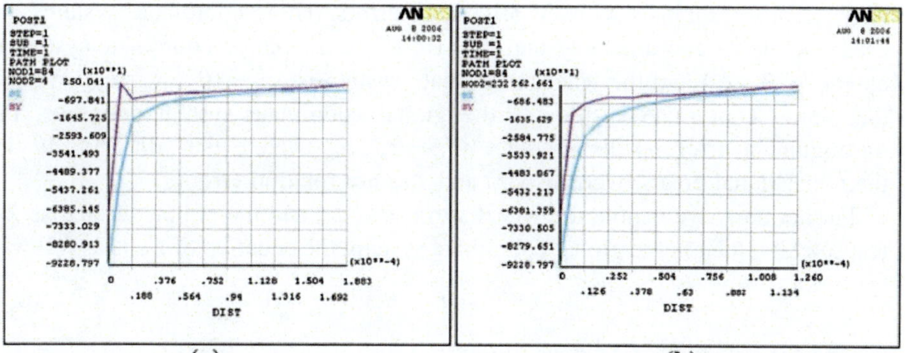

(a) (b)

Fig. 6.14 The stress distribution along depth. **a** The wheel, **b** The pinion

Table 6.5 Maximum normal stress in the contact zone (in MPa)

	Single-tooth model	Whole gear model	Difference between the two models (%)
Center wheel—third pinion	−0.983	−0.996	1.32
Third wheel—forth pinion	−0.997	−0.986	1.10

Table 6.6 Maximum shear stress in the contact zone (in MPa)

	Single-tooth model	Whole gear model	Difference between the two models (%)
Center wheel–third pinion	0.975	1.358	39.28
Third wheel–forth pinion	0.897	0.927	3.34

center wheel is greater than that of the third wheel and hence, the error is greater. In this case, the whole model should be used.

In addition, Tables 6.5 and 6.6 show that the maximum normal stress and shear stress of two gear pairs are both significantly lower than the yield stress of the materials (steel and brass). Therefore, the gears shall not fail in the movement and torque transmission.

6.4 Misalignment Errors and Their Effects

6.4.1 The Description of the Misalignment Errors

During the manufacturing and assembling of gears, errors are inevitable. These errors all transform to misalignment errors and affect the timekeeping accuracy. In general, the misalignment errors can be classified into the translational error and the rotational error. As an example, Fig. 6.15 shows the barrel wheel and the center pinion gear pair with misalignment errors. Three coordinate systems are used: the global coordinate system $\sigma = (\theta, R, Z)$, the pinion coordinate system is $\sigma_p = (\theta_p, R_p, Z_p)$ and the wheel coordinate system is $\sigma_w = (\theta_w, R_w, Z_w)$. We use the pinion coordinate system as the global coordinate system and thus, the misalignment error results in $\sigma_{w'} = (\theta_{w'}, R_{w'}, Z_{w'})$. Note that ΔR and $\Delta\theta$ are the translational errors, whereas Δr_θ and Δr_R are rotation errors.

Furthermore, we denote the wheel surface as Σ_w and the pinion surface as Σ_p, respectively; then in the global coordinate system, the equation of Σ_w is as follows:

$$r^{(w)} = r^{(w)}(u, v) \tag{6.7}$$

And the equation of Σ_p is as follows:

$$\begin{cases} r^{(p)} = M_{pw} r^{(w)} \\ n \cdot v^{(wp)} = 0 \end{cases} \tag{6.8}$$

where n is the unit normal vector, $v^{(wp)}$ is the relative velocity in the contact point between Σ_w and Σ_p and M_{pw} is the coordinate transform matrix from σ_w to σ_p. Normally, the coordinate transform matrix can be written as follows:

$$M_{pw} = \begin{bmatrix} 1 & 0 & 0 & 0 \\ 0 & 1 & 0 & R_p + R_w \\ 0 & 0 & 1 & 0 \\ 0 & 0 & 0 & 1 \end{bmatrix}$$

where R_p and R_w are the radians of the pinion and the wheel, respectively.

When the misalignment errors occur, the equation $\Sigma_{w'}$ becomes:

$$r^{(w)\prime}(u, v) = M_e r^{(w)}(u, v) \tag{6.9}$$

where $M_e = M_e^1 \cdot M_e^2 \cdot M_e^3$

$$M_e^1 = \begin{bmatrix} 1 & 0 & 0 & \Delta\theta \\ 0 & 1 & 0 & \Delta R \\ 0 & 0 & 1 & 0 \\ 0 & 0 & 0 & 1 \end{bmatrix}$$

$$M_e^2 = \begin{bmatrix} 1 & 0 & 0 & 0 \\ 0 & \cos\Delta r_\theta & -\sin\Delta r_\theta & 0 \\ 0 & \sin_\theta & \cos\Delta r_\theta & 0 \\ 0 & 0 & 0 & 1 \end{bmatrix}$$

$$M_e^3 = \begin{bmatrix} \cos\Delta r_R & 0 & -\sin\Delta r_R & 0 \\ 0 & 1 & 0 & 0 \\ \sin\Delta r_R & 0 & \cos\Delta r_R & 0 \\ 0 & 0 & 0 & 1 \end{bmatrix}$$

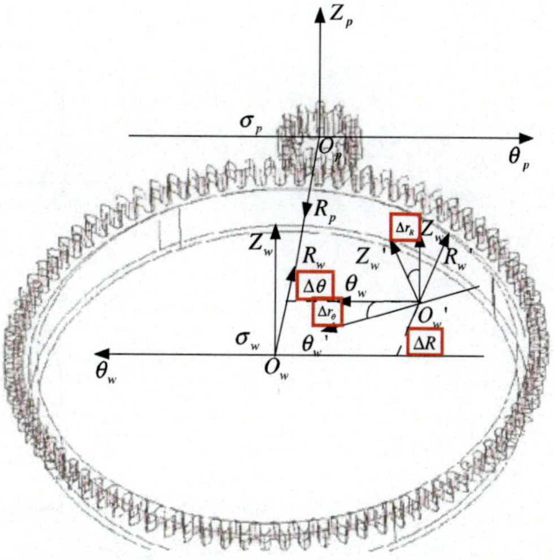

Fig. 6.15 Illustration of the misalignment error

The effect of the misalignment error can be analyzed using the loaded tooth contact analysis (LTCA).

6.4.2 Loaded Tooth Contact Analysis

The objective of LTCA is to find the contact zone and the displacement of the conjugated teeth under loading. We use the surface-to-surface searching algorithm to solve this elastic contact problem. This algorithm does not require pre-designed contact pattern and contact zone and hence, is effective. The flow chart of the algorithm is shown in Fig. 6.16.

From the mathematical point of view, before the two gears coming into contact, the static potential energy function of the contact system is:

$$\Pi = \frac{1}{2}\mathbf{U}^T\mathbf{K}\mathbf{U} - \mathbf{U}^T\mathbf{P} \tag{6.10}$$

where **U** is the displacement matrix, **K** is the stiffness matrix and **P** is the external force acting on the system. On the contact surfaces, the displacement must satisfy the non-penetration equation:

$$\mathbf{G} = \mathbf{u}_1^n - \mathbf{u}_2^n - \mathbf{g} \leq 0 \tag{6.11}$$

where **G** is the distance between two contact surfaces and **g** is the initial distance between the two contact surfaces. On the contact, the static potential energy function of the contact system changes to:

$$\Pi = \frac{1}{2}\mathbf{U}^T\mathbf{K}\mathbf{U} - \mathbf{U}^T\mathbf{P} + \mathbf{R}_n^T\mathbf{G} + \frac{1}{2}\mathbf{G}^T\gamma\mathbf{G} \tag{6.12}$$

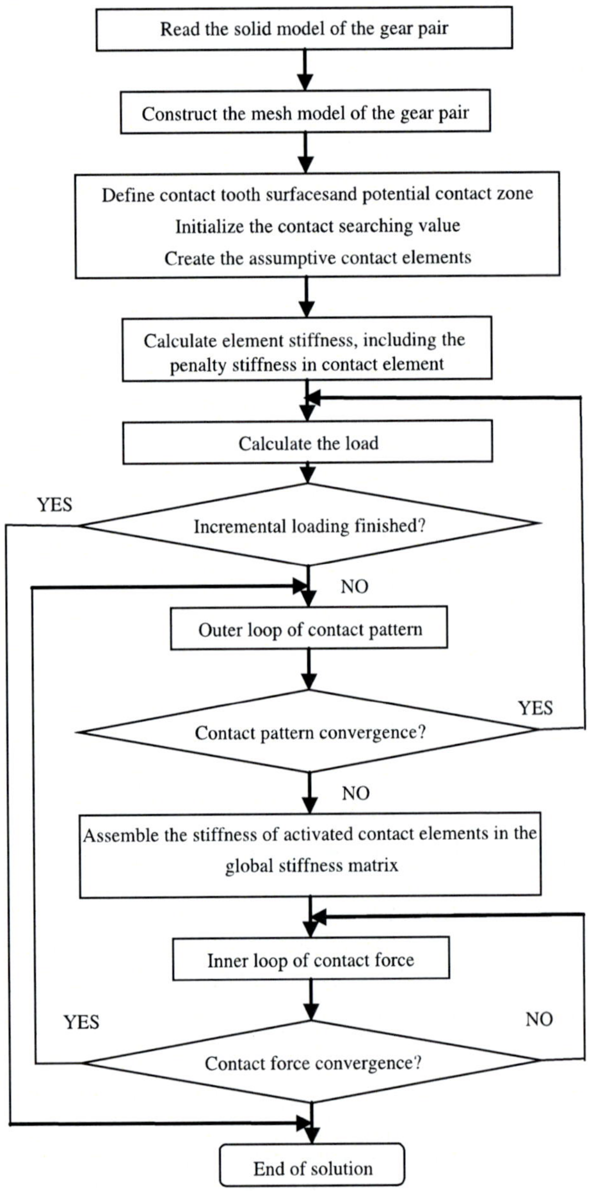

Fig. 6.16 The flow chart of the LTCA

where $\mathbf{R}_n$ is the normal contact force matrix and γ is a penalty number. Discretize Eq. (6.11), we have:

$$\mathbf{G} = \sum_{i=1}^{nH} \mathbf{N}_H^i \mathbf{u}_H^{in} - \sum_{i=1}^{nT} \mathbf{N}_T^i \mathbf{u}_T^{in} - \mathbf{g} \tag{6.13}$$

where **N** is the nodal function. This equation can be rewritten in the matrix form:

$$\mathbf{G} = \mathbf{N}\mathbf{U} - \mathbf{g} \tag{6.14}$$

Substitute Eq. (6.14) into Eq. (6.12), it follows that:

$$\Pi = \frac{1}{2}\mathbf{U}^T\mathbf{K}\mathbf{U} - \mathbf{U}^T\mathbf{P} + \mathbf{R}_n^T(\mathbf{N}\mathbf{U} - \mathbf{g}) + \frac{1}{2}(\mathbf{N}\mathbf{U} - \mathbf{g})^T\gamma(\mathbf{N}\mathbf{U} - \mathbf{g}) \tag{6.15}$$

Based on the minimum energy principle, $\frac{\partial \Pi}{\partial \mathbf{U}} = 0$, we have the following:

$$\left(\mathbf{K} + \mathbf{N}^T\gamma\mathbf{N}\right)\mathbf{U} = \mathbf{P} - \mathbf{N}^T(\mathbf{R}_n + \gamma\mathbf{g}) \tag{6.16}$$

On the other hand, the normal restraint that the contact element must satisfy is:

$$\mathbf{G} \times \mathbf{R}_n = 0 \tag{6.17}$$

Hence, the possible contact elements could only be activated when $\mathbf{G} = 0$. Note that the searching procedure only needs to calculate the contact pressure of the activated elements.

Furthermore, when the friction is considered, additional constraint conditions are applied onto the activated contact elements:

$$\begin{aligned} &\xi = |\mathbf{R}_t| - \mu\mathbf{R}_n \leq 0 \pm \cdots \\ &\Delta\mathbf{U}^t = \Delta\tau\frac{\mathbf{R}_t}{|\mathbf{R}_t|} \\ &\xi \times \Delta\tau = 0 \end{aligned} \tag{6.18}$$

where ξ is the friction constrain, $\mathbf{R}_t$ is the contact friction force matrix and $\Delta\tau$ is the magnitude of $\Delta\mathbf{U}^t$. The friction contact constraint implies that if there is a slipping between the two surfaces ($\Delta\tau > 0$), then $\xi = 0$; and if the two surfaces are sticking together ($\Delta\tau = 0$), then $\mathbf{R}_t = 0$.

Using Eqs. (6.16) and (6.17), the normal contact force $\mathbf{R}_n$ can be computed iteratively:

$$\mathbf{R}_n^i = \mathbf{R}_n^{i-1} + \gamma\mathbf{G} \tag{6.19}$$

Note that in the iterations, Eq. (6.14) should be checked. The iteration is converged when $\mathbf{G} = 0$. Furthermore, the contact friction force $\mathbf{R}_t$ can be calculated using Eq. (6.18). From a physical point of view, $\mathbf{R}_n$ represents the transmitted torque, whereas $\mathbf{R}_t$ is related to the energy loss due to the friction.

6.4.3 The Effect of the Misalignment Errors on the Contact Zone

Using the LTCA, the effect of the misalignment errors can be simulated. The misalignment errors can be classified into translational error and rotational error.

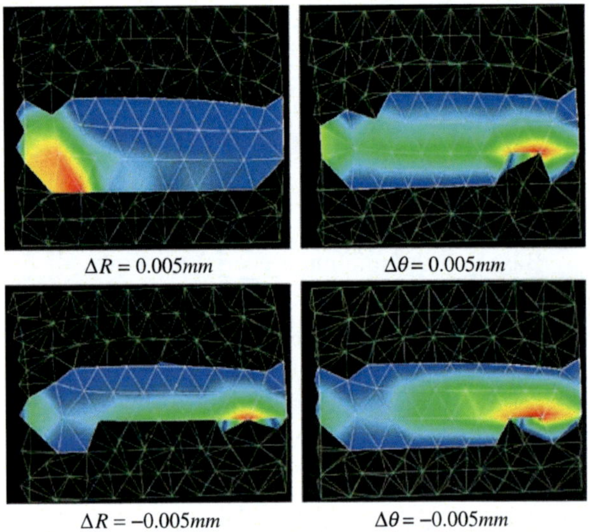

Fig. 6.17 The effect of the translational error on the contact zone of the pinion

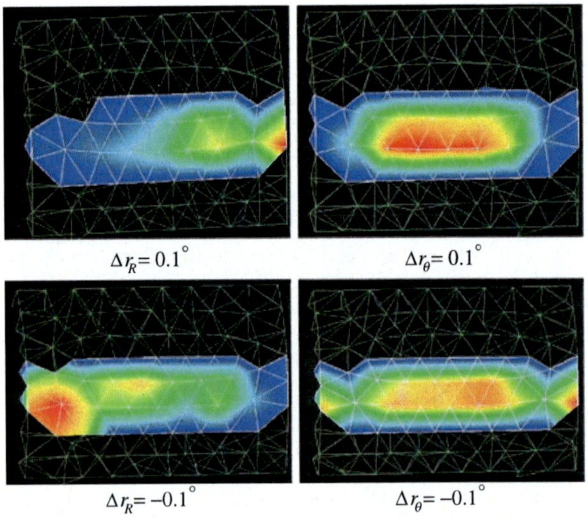

Fig. 6.18 The effect of the rotational errors on the contact zone of the pinion

The simulation results of the pinion are shown in Figs. 6.17 and 6.18, respectively. From the figures, it is seen that when the misalignment errors occur, the contact zone shifts to the edge of the tooth. In addition, the translational errors would have larger effect than that of the rotational errors because the translational errors are more direct. In particularly, under the presence of the misalignment error Δr_θ, the contact zone is still in the middle of the tooth surface. The simulation results of the wheel are rather similar.

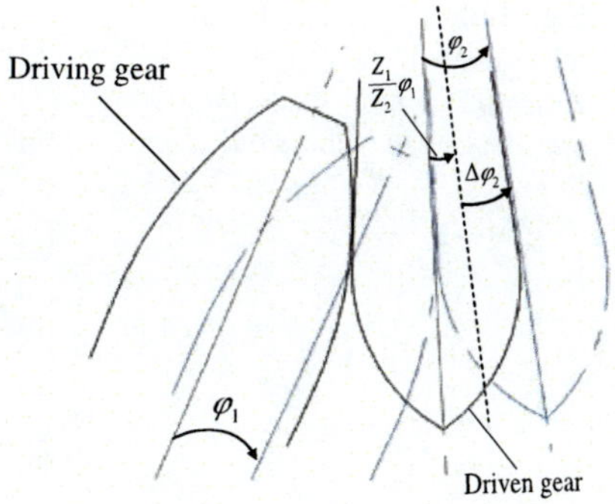

Fig. 6.19 Illustration of the transmission error

6.4.4 Misalignment and Transmission Error

Transmission error (TE) is caused by the misalignment errors and the tooth deformation. As shown in Fig. 6.19, the misalignment-induced error can be described by the following equations:

$$\Delta\varphi_2 = \varphi_2 - \frac{Z_1}{Z_2}\varphi_1 \tag{6.20}$$

$$\varphi_2 = \frac{\frac{F}{K} + \delta H + \delta W + \delta I}{r_2} \tag{6.21}$$

where φ_1 is the rotation angle of the driving gear; φ_2 is the rotation angle of the driven gear; δH, δW and δI are the displacement due to the tooth bending, the tooth shearing and the initial tooth distance, respectively; Z_1 and Z_2 are the tooth number of the driving gear and the driven gear, respectively; K is the tooth stiffness; and F is the load applied on meshing tooth pair.

TE can be computed using LTCA. We use the barrel wheel (driving gear) and the center pinion (driven gear) as an example. The center pinion has 17 teeth and hence, its meshing circle covers 21.17 degrees. The LTCA is executed every time when the pinion is being rotated 0.5 degree and thus, there are a total of 43 simulation steps. Figure 6.20 shows two simulation examples: With the translational error $\varDelta R = 0.005$ mm, TE is always positive (the *dash line* in the figure), whereas with the rotational error $\varDelta r_R = 0.1°$, TE varies (the *solid line*).

The deformation-induced TE can be determined by the following equation:

$$\psi = L_1\Delta\varphi = L_1 L_2 F \tag{6.22}$$

where L_1 and L_2 are the interference coefficients and F is the load applied onto the meshing tooth pair, which can be calculated as follows.

$$F = \sum_{i=1}^{m} F_i\Delta s_i = \sum_{i=1}^{m} \frac{\Delta\varphi_i}{L_2(i)}\Delta s_i = \int_A \mathbf{R}_n ds = \sum_{i=1}^{m} \mathbf{R}_{ni}\Delta s_i \tag{6.23}$$

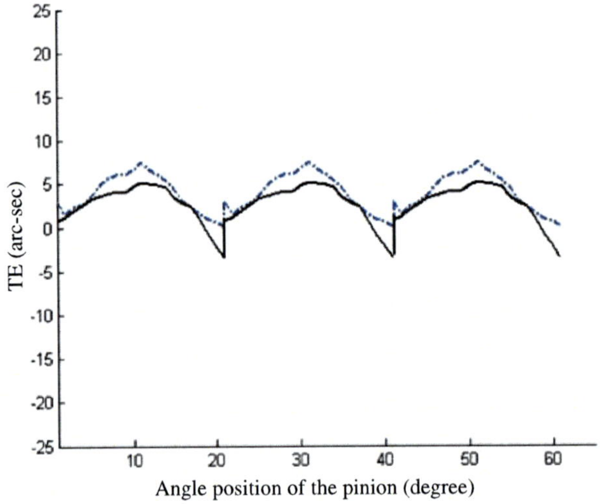

Fig. 6.20 Two examples of misalignment induced TE: $\Delta R = 0.005$ mm (*dash line*) and $\Delta r_R = 0.1°$ (the *solid line*)

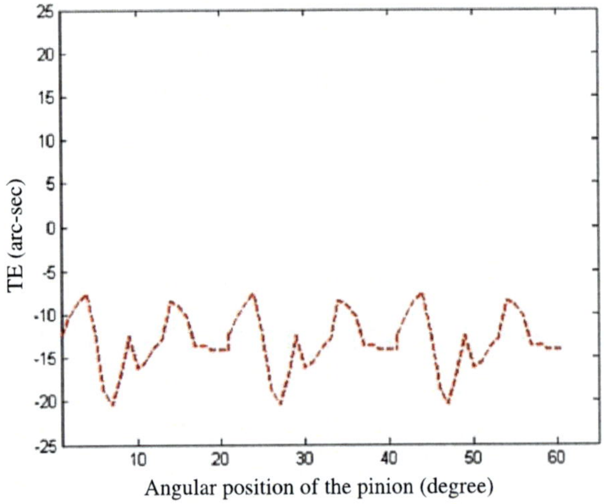

Fig. 6.21 An example of the tooth deformation induced TE

Figure 6.21 shows an example of the deformation-induced TE. Note that when a gear pair comes into contact, the driven gear (the center pinion) is assumed fixed. As the driving gear (the barrel wheel) pushes the driven gear, the driven gear is deformed and hence, TE is negative.

6.4.5 Misalignment and Torque Transmission

Misalignment affects torque transmission as well. The calculation of torque transmission is also based on LTCA. As discussed in Sect. 6.4.2, LTCA gives the contact pressure $\mathbf{R}_n$; accordingly, the torque transmission can be found using the formula below:

Fig. 6.22 Illustration of the torque transmission calculation

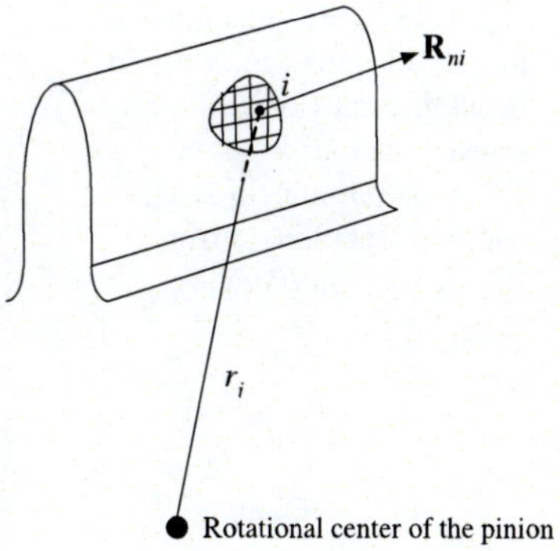

$$T = \sum_{i=1}^{m} \mathbf{R}_{ni} r_i \qquad (6.24)$$

where T is the torque transmitted in the gear pair, $\mathbf{R}_{ni}$ is the contact pressure of the ith element and r_i is the radian of the ith element. This is illustrated in Fig. 6.22.

Figure 6.23 shows three examples: with no misalignment (in *red*), with translational error $\Delta R = 0.005$ mm (in *blue*) and with rotational error $\Delta r_R = 0.1°$ (in *black*). From the figure, it is seen that without misalignment error, the variation of the torque is usually within 10%. With the rotational error, the variation becomes 87%, and with the translational error, the variation reaches 145%. In addition, with misalignment errors, the pattern of the torque transmission changes as well, causing the edge contact. In other words, the misalignment errors would bring down the smoothness of the torque transmission and hence, reduce the timekeeping accuracy. To solve this problem, we propose to modify the tooth profile (Xu et al. 2007), which will be detailed in the subsequent section.

6.5 Tooth Profile Modification

As discussed above, the misalignment errors will bring down the timekeeping accuracy. Though, it is well known that the misalignment errors are inevitable in manufacturing and assembling. To minimize the effect of the misalignment errors, we therefore proposed to modify the tooth profile (Xu et al. 2007).

6.5.1 The Modified Tooth Profile

The objectives of the tooth profile modification are as follows:

(1) Decrease the tooth edge contact.
(2) Reduce the variation of torque transmission.
(3) Reduce TE.

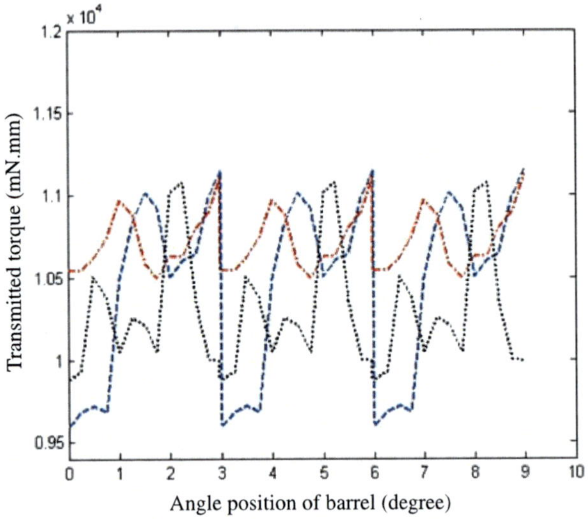

Fig. 6.23 Three examples of torque transmission: with no misalignment (in *red*), with translational error $\Delta R = 0.005$ mm (in *blue*), and with rotational error $\Delta r_R = 0.1°$ (in *black*)

A careful examination of the tooth profile (Fig. 6.3) reveals that the sharp addendum of the pinion tooth plays little role in the contract and the torque transmission. Therefore, it can be cut down as shown in Fig. 6.24a. This makes the meshing entry angle smaller and hence, decreases the energy loss due to tooth contact friction. In addition, as shown in Fig. 6.24a, the modified tooth is a slightly expanded in the middle. This has the effect of slowing down the rotation of the gear on contact, which gives a positive TE and reduces the total TE. Ideally, the aforementioned tool profile modification would only apply in the middle to accommodate rotational misalignment errors. However, for the easy of manufacturing, the edge of the tooth shall be straight, as shown in Fig. 6.24b.

On defining the modified features, the skinning method (Shi 1993) is utilized to complete the definition of the tooth profile by interpolation. We use the B-Spline curve for interpolation. Note the interpolation is not necessary to go through the control point lattice, but through a series of section curves. Before the interpolation, all the section curves should be adjusted to have a same degree-of-freedom, k, and nodal vector $\mathbf{V} = [v_0, v_1, \cdots, v_{m+2k}]$. The section curves are defined as follows:

$$s_j(v) = \sum_{i=0}^{m+k+1} d_{i,j} N_{i,k}(v) \qquad j = 0, 1, \cdots n \tag{6.25}$$

where d_{ij} are the control points and $N_{ik}(v)$ is the base function of the B-Spline curve.

Then, ridgelines are specified along the longitudinal direction of the tooth. Finally, combining all section curves along the ridgelines, the modified tooth profile is constructed.

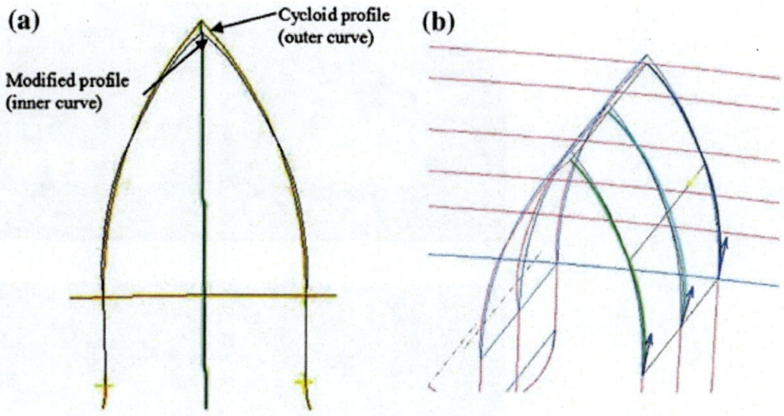

Fig. 6.24 Illustration of the tooth profile modification of a pinion. **a** 2D view **b** 3D view

Fig. 6.25 The effect of translational errors on the contact zone after the tool profile modification

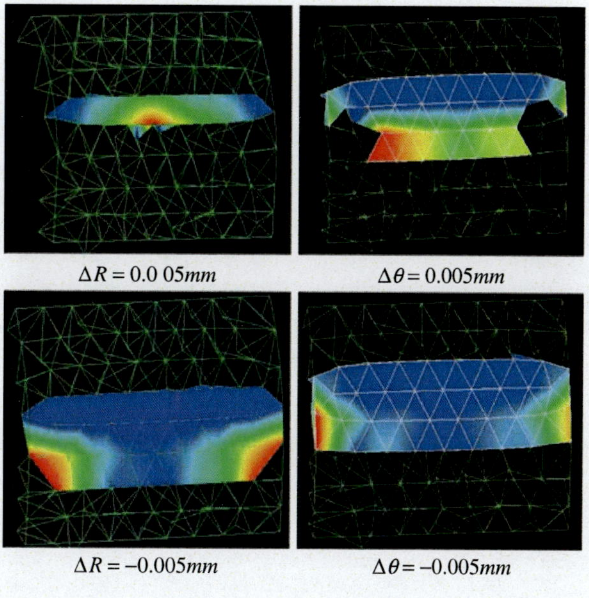

6.5.2 Tooth Profile Modification and the Contact Zone

Using the LTCA, the tooth contact zone can be computed. Figures 6.25 and 6.26 show the effect of translational errors and rotational errors on the contact zone after the tooth profile modification. In comparison to Figs. 6.16 and 6.17, it is seen that the contact zones are now symmetrically located along the tooth width, which implies the effects of the misalignment errors have been reduced.

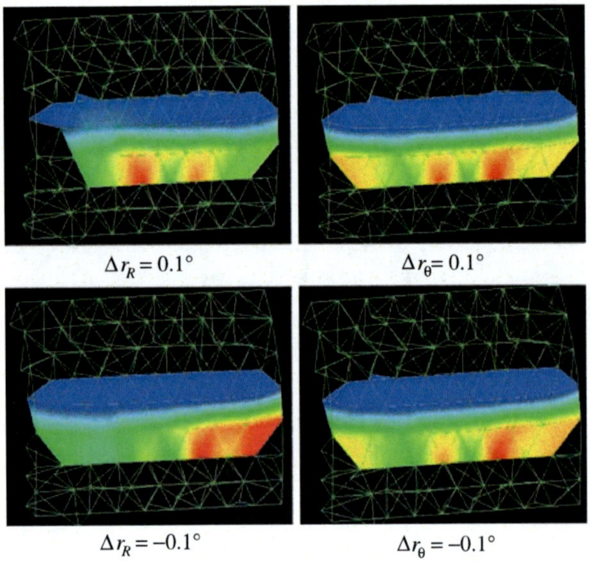

Fig. 6.26 The effect of rotational errors on the contact zone after the tool profile modification

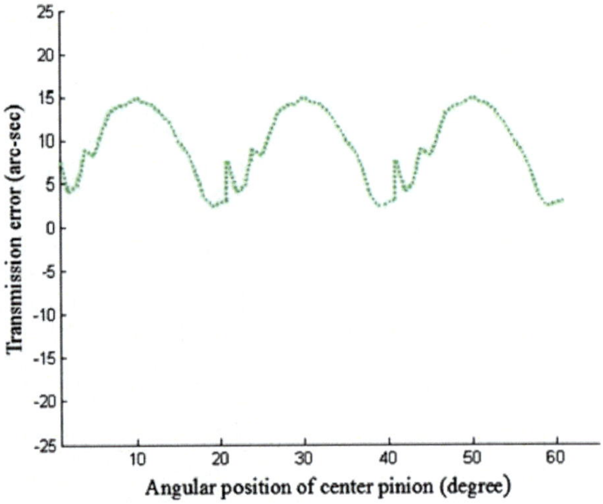

Fig. 6.27 Transmission error due to tooth modification

6.5.3 Tooth Profile Modification and TE

As mentioned before, the modified wheel tooth profile is a slightly "fatter" curve than cycloid, which will introduce the positive TE to the gear pair. Figure 6.27 shows the TE due to wheel tooth modification. Figures 6.28 and 6.29 illustrate the comparison of TE with and without wheel tooth modification when translational error $\Delta R = 0.005$ *mm* and rotary error $\Delta r_R = 0.1°$ exist, respectively. In Fig. 6.26, the Maximum magnitude of TE is 11.22 arc-sec without wheel tooth modification

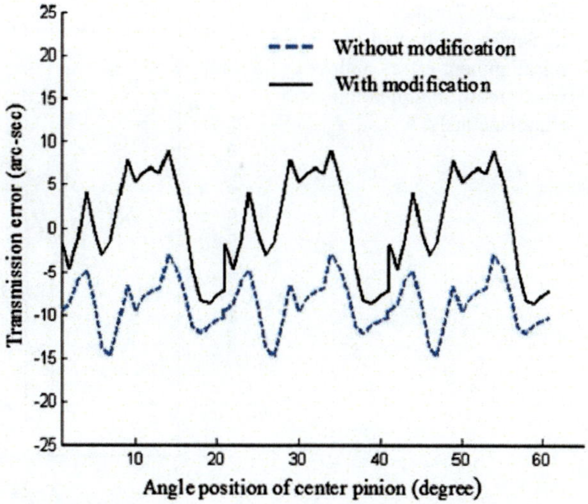

Fig. 6.28 Comparison of TE with and without wheel tooth modification when translational error $\Delta R = 0.005$ mm exists

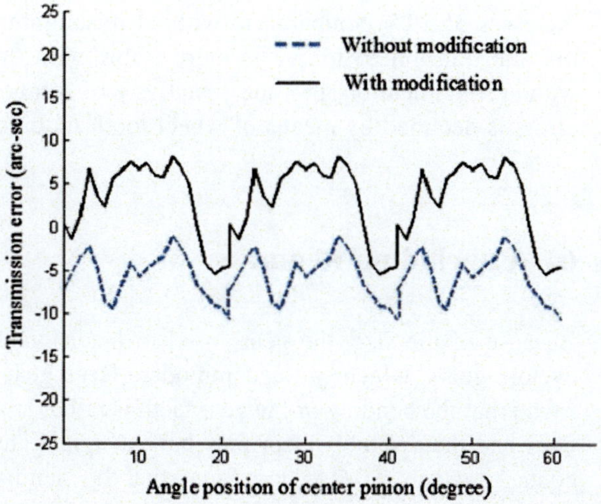

Fig. 6.29 Comparison of TE with and without wheel tooth modification when rotary error $\Delta r_R = 0.1^\circ$ exists

and 8.76 arc-sec with modification. And in Fig. 6.20, the Maximum magnitude of TE is 8.39 arc-sec without wheel tooth modification and 7.07 arc-sec with modification.

6.5.4 Torque Transmission After Tooth Modification

With the influence of misalignment errors and wheel tooth modification simultaneously, the torque transmission of barrel wheel and center pinion is indicated in Fig. 6.30. From the figure, it is seen that the fluctuation of transmitted torque

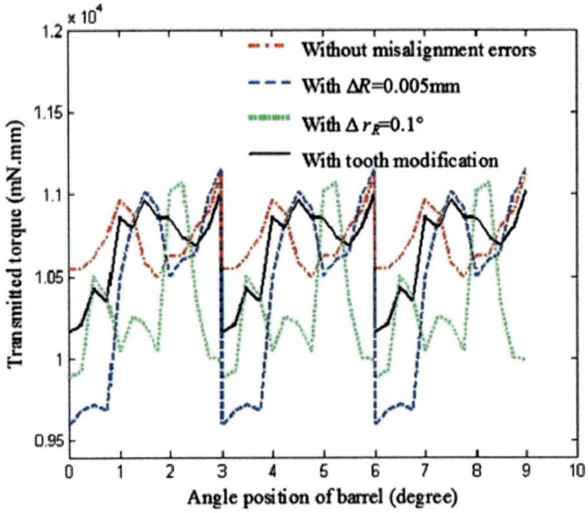

Fig. 6.30 Torque transmission affected by misalignment errors and wheel tooth modification simultaneously

increases 35.29% compared to without misalignment errors. However, it is clear that the transmitted torque is more stable with the tooth modification than that without modification, i.e., the sensitivity of torque transmission to misalignment errors is declined by means of wheel tooth modification.

6.6 Concluding Remarks

Because of the size, the gears used in mechanical watch movement are usually cycloid gears, which can accommodate large gear ratio. Based on the FEA, it is found that the bending of the gear teeth is rather small. Though, the manufacturing error and the assembly error may have a significant effect. Using the loaded tooth contact analysis (LTCA), it is found that the manufacturing error and the assembly error would cause significant transmission error (TE) and torque variation. To reduce the TE and the torque variation, one can modify the tooth profile.

References

Bair BW, Tsay CB (1998) ZK-type dual-lead worm and worm gear drives. J Mech Des 120:414–421

Bibel G (2011) Procedure for tooth contact analysis of a face gear meshing with a spur gear using finite element analysis. Available at http://glltrc.nasa.gov/GLTRS. Jan 2001, Accessed 12 Dec 2011

Bibel GD, Reddy SK, Savage M, Handschuh RF (1994) Effects of rim thickness on spur gear bending stress. J Mech Des 116:1157–1162

Feng PH, Litvin FL (1998) Determination of principal curvatures and contact ellipse for profile crowned helical gears. In: Proceedings of 1998 ASME design engineering technical conferences, Atlanta, USA, 13–16 Sept 1998

Fong Z, Tsay C (1991) A mathematical model for the tooth geometry of circular-cut spiral bevel gears. J Mech Des 113:174–181

Greening JH, Barlow RJ, Loveless WG (1980) Load sharing on the teeth of double enveloping worm gear. In: Power transmission and gearing conference, paper No. 80-C2/DET-43, San Francisco, U S A

Litvin FL, Fuentes A (2004) Gear geometry and applied theory. Cambridge University Press, USA

Litvin FL, Lu J (1995) Computerized design and generation of low-noise double circular-arc helical gears with low transmission errors. Comput Methods Appl Mech Eng 127:57–86

Litvin FL, Chen NX, Lu J, Handschuh RF (1995) Computerized design and generation of low-noise helical Gears with modified surface topology. J Mech Des 117:254–261

Shi FZ (1993) CAGD&NURBS. Beijing Aeronautics and Astronautics University Publishing Company, China, p 463

Simon V (1993) Stress analysis in double enveloping worm gears by finite element method. J Mech Des 115:179–185

Soel IH, Litvin FL (1996) Computerized design, generation and simulation of meshing and contact of worm-gear drives with improved geometry. Comput Methods Appl Mech Eng 138:73–103

Varangan S, Ganesan N (1994) Static contact stress analysis of a spur gear tooth using the finite element method. Comput Struct 51:765–770

Wei ZP (2011) Stress and deformations in involute spur gears by finite element method, M.Sc. thesis of the U. of Saskatchewan. Available at http://library.usask.ca/theses/available/etd-10272004-163322/unrestricted/ThesisFinalVersion.pdf. Accessed 12 Dec 2011

Wikipedia (2002) Gear. http://en.wikipedia.org/wiki/Gear. Accessed 12 Dec 2011

Wild JM (2003) Wheel and pinion cutting in horology: a historical and practical guide. The Crowood Press Ltd. England

Xu WJ, Qin DT, Fu Y, Du R (2007) A new approach for reducing the cycloid gear's sensitivity to misalignment errors in precision transmission. In: Conference on integration and commercialization of micro and nano systems, Sanya, China, 13–17 Jan 2007

Chapter 7
Concluding Remarks

Mechanical watches and clocks are arguably the most complex of "pure" mechanical devices. Sure, the combustion engine is complex, but its complexity lays in the combustion and its mechanism is just a cam mechanism. The crane towers is complex, but their complexity lays in its size and its mechanism is just a three-bar mechanism. Robots are complex, but its complexity comes from their controls and a robot's mechanism is just an assembly of links and joints.

However, the escapement of the mechanical watch is truly complex. First, it is made of three components: the escape wheel, the pallet fork and the balance wheel with the hairspring. These components are driven by an external torque as well as the internal reaction force of the hairspring. They must work in perfect harmony to carry out timekeeping, which requires absolute precision. Second, the motion of the escapement is determined by a series of impacts among the escape wheel, the pallet fork and the balance wheel. These impacts are necessary for power transportation. However, they also cause change in force and motion (displacement, velocity and acceleration). The effects of these impacts are complex. In fact, the complexity of the impact mechanics is well known to be difficult to solve. Third, from a practical point of view, the motion of the escapement involves many factors, such as friction, temperature, magnetics, manufacturing error, assembly error and material fatigue. Any one of these factors can easily jeopardize the accuracy of timekeeping. Today, the design and manufacturing of mechanical watches and clocks is still considered an art, as well as an advanced use of science and technology.

In general, making a mechanical watch involves the following aspects: (a) design, (b) design analysis, (c) materials and processing (e.g., heat treatment and surface polishing) (d) manufacturing, (e) assembly and (f) quality control. This monograph covers only a portion of part (b). We wish to share with the readers some of our thoughts on the other aspects herein.

(a) Design. The brain of the mechanical watch and clock is the escapement. At present, 99% of mechanical watches use the Swiss lever escapement. The rest of

R. Du and L. Xie, *The Mechanics of Mechanical Watches and Clocks*,
History of Mechanism and Machine Science 21, DOI: 10.1007/978-3-642-29308-5_7,

the design serves only to put everything in order. Throughout the years, some "fundamental" designs have been developed and proved to be effective. Based on these designs, one can add functions such as date, calendar, moon phase, stop-watch, Tourbillion, etc. There is little need to create a new design. Recently, two new escapements, the Daniel co-axial escapement and the Ulysee Nardin dual wheel escapement, were developed. From a mechanics point of view, they are better in terms of force transmission and motion regulation. However, they are also more difficult to make. We expect that more new escapements will be invented in the near future. For these new inventions, the key issue will be how to transport power to the balance wheel while not interrupting its rhythm. In addition, manufacturing and assembly must be considered.

(b) Design analysis. As mentioned earlier, this monograph tries to give an in-depth view of the mechanical watch and clock from the context of modern mathematics. It covers the Swiss lever escapement (Chap. 3), the spiral spring (Chap. 4), the automatic winding device (Chap. 5) and the gear train (Chap. 6). We wish to emphasize the importance of mathematical modeling and computer simulation, as it helps to get the design right the first time. Throughout the years, designs were realized by trial and error, which is very time-consuming. We must take the advantage of the modern computer technology to make the design more effective and efficient.

(c) Materials and processing. Traditionally, most, if not all, of the components of a mechanical watch and clock we made of steel or brass. In recent years, new materials have been adopted, such as silicon for the hairspring and bridges as well as sapphire for the main plates. Furthermore, the material-processing techniques have gotten more advanced. In particular, the diamond like coating (DLC) is noteworthy. According to current literature, DLC can create surfaces with minimum friction, making it very attractive for watch and clock production. It has even been considered for replacing the ruby bearings. However, ruby bearings are also used to adjust the heights of the components and hence, may be irreplaceable.

(d) Manufacturing. In general, the components in a mechanical watch and clock must be accurate to within a few or a few dozen micrometers, a size less than a human hair. Clearly, such accuracy is difficult to achieve. Nowadays, with sensors and computer control technology, it is possible to hold such micrometer accuracy in production. In our experience, a cost-effective way is to work on the tools, fixtures and dies. For example, mainplates of quartz watches are currently being made by forming (a combination of forging and stamping). Is it possible to make the mainplate of the mechanical watch by the same method? This would cut the production time many fold.

(e) Assembly. Assembly of watches still depends greatly on skilled craftsmen. Automated assembly is not only difficult (because of the accuracy requirement) but also unnecessary (because of the quantity). Similar to manufacturing, an effective way to improve productivity is to work on the tools and fixtures. Many small innovative devices can help to cut the cost and improve the quality.

(f) As a luxury product, mechanical watches and clocks must have excellent quality and hence, quality control (QC) is very important. QC must be applied to each and every step of the production: the incoming materials and components, the manufactured components, the subassemblies and the final watch produced. One may also add modern industrial engineering QC tools, such as failure mode and effect analysis (FMEA), Fault Tree Diagnosis and Six Sigma. Our experience has taught us that the QC is the most important part of watch and clock making, especially when launching a new product.

Last but not the least we hope that the materials presented in this monograph will promote more innovation in the watch and clock industry and other industries. It has been shown that those who master the technologies of watch manufacturing have advantages in other industries as well, including machining, metal forming, tribology, etc. In short, the watch industry should not be overlooked.

Appendix
Computer Animation

Please visit the website http://extra.springer.com/ to download

The Verge Escapement
The Graham Escapement
The Grasshopper Escapement
The Cylinder Escapement
The Spring Detent Escapement
The English Lever Escapement
The Swiss Lever Escapement
The Daniels' Co-Axial Escapement
The Independent Double Wheel Escapement
The Double Wheel Escapement
The ETA automatic winding device
The Seiko automatic winding device

R. Du and L. Xie, *The Mechanics of Mechanical Watches and Clocks*,
History of Mechanism and Machine Science 21, DOI: 10.1007/978-3-642-29308-5,

Index

R. Du and L. Xie, *The Mechanics of Mechanical Watches and Clocks*,
History of Mechanism and Machine Science 21, DOI: 10.1007/978-3-642-29308-5,